T0093681

# PERIODIC TABLE EXPLORER

ADRIAN DINGLE

San Diego, California

**Thunder Bay Press**
An imprint of Printers Row Publishing Group
9717 Pacific Heights Blvd, San Diego, CA 92121
www.thunderbaybooks.com • mail@thunderbaybooks.com

First published as The Elements in 2017, this edition first published in 2022


Correspondence regarding the content of this book should be sent to Thunder Bay Press, Editorial Department, at the above address. Author and image inquiries should be addressed to The Bright Press at the address below.

**Thunder Bay Press**
Publisher: Peter Norton
Associate Publisher: Ana Parker
Editor: Dan Mansfield
Acquisitions Editor: Kathryn Chipinka Dalby

This book was conceived, designed, and produced by
The Bright Press, an imprint of The Quarto Group
The Old Brewery, 6 Blundell Street,
London N7 9BH
United Kingdom
T (0)20 7700 6700
www.Quarto.com

**The Bright Press**
Publisher: James Evans
Editorial Director: Isheeta Mustafi
Managing Editor: Jacqui Sayers
Art Director: James Lawrence
Project Editors: Joanna Bentley, Nigel Matheson
Designer: Tony Seddon
Scientific Consultant: Dr. Kit Chapman

ISBN: 978-1-64517-941-2

Cover photo: Shutterstock/Romolo Tavani

Printed in China

26 25 24 23 22 1 2 3 4 5

# Contents

# Introduction

**"Stuff," or more scientifically speaking, matter, is everywhere. If something takes up space and has mass, then it is matter. That's a pretty diverse group of things that, on the face of it, don't appear to have much in common with one another—however, they are much more alike than you may imagine. All matter is composed of some basic building blocks; more precisely, 118 building blocks (that we know of), which when arranged in an infinite number of ways produce that matter.**

The history of these building blocks is as diverse as their properties, ranging from discovery stories that are lost in the mist of ancient times, to ones that have only been officially confirmed and named very recently. They also have an incredibly diverse set of characteristics (physical and chemical properties) of their own, each with a unique place in the grand scheme. What are these building blocks that we speak of? They are more commonly known as the elements.

## KEY FIGURE

### DMITRI MENDELEEV 1834–1907

The Russian chemist Dmitri Mendeleev is the pivotal figure in the history of the periodic table, as he was the first to propose an organization of elements that resembles the modern-day table. In truth, important work by many other scientists had preceded Mendeleev, but the Russian's ordering of all of the elements by atomic weights revealed patterns that are still relevant and familiar today. His table of 1869 also left gaps that allowed for the prediction of, at the time, undiscovered elements.

### ORDERING THE ELEMENTS

The 118 elements are corralled and called to order in one of the most iconic symbols of science of all time: the periodic table. Whole books have been devoted to the history and development of the table, and like almost all scientific endeavors, its evolution can be attributed to the work of many people. The person usually considered the "father" of the periodic table is the Russian chemist Dmitri Mendeleev (1834–1907), but many others preceded his genius. Similarly, since Mendeleev's first table of 1869, many more have built upon his work to get us to the modern IUPAC (International Union of Pure and Applied Chemistry) table that we use today. Mendeleev's successors include those who spearheaded the development of the atomic model in the late nineteenth and early twentieth centuries, such as Dane Niels Bohr (1885–1962), Austrian Erwin Schrödinger (1887–1961), Englishman J. J. Thomson (1856–1940), New Zealander Ernest Rutherford (1871–1937), Englishman James Chadwick (1891–1974), and Frenchman Louis de Broglie (1892–1987), along with incredibly important (but generally less well known) contributions from many others. Henry Moseley (1887–1915) is one such person, and his contribution marked a profound change in terms of the way that the elements on the periodic table were ordered.

### DISCOVERING THE ATOMIC NUMBER

In Mendeleev's original table the elements were ordered by their atomic weights. This was an idea that persisted with him, to the point of his "explaining away" some apparent anomalies. For example, in Mendeleev's table tellurium and iodine were "incorrectly" ordered, with the heavier Te (number 52) coming before the lighter I (number 53) in his scheme. At the time, Mendeleev said that the atomic weights of either one (or both) of these elements had been incorrectly calculated, thus explaining the break in his sequence.

It took experiments by Moseley in 1913—building on earlier work by Anton van den Broek (1870–1926)—to make more sense of the apparent anomaly caused by Mendeleev's ordering. Moseley determined that elements, when struck with an incident energy source, emitted X-rays. When he conducted the same experiment on fourteen of the elements known at the time, he found a mathematical relationship between the X-rays and the square of an integer that represented the element's position on the table. What Moseley had done was to discover what we know now as the atomic number, and it is by atomic number that the periodic table is now ordered.

Moseley's work also had at least one other important impact on the periodic table. Since his experiments accurately determined the positions of the known elements, they also allowed chemists to see that elements were missing from the sequence, thus allowing them to target their searches for new elements more specifically. These targeted searches led to the discovery of seven "missing" elements within the first ninety-two, namely those with atomic numbers of 43, 61, 72, 75, 85, 87, and 91. Promethium, number 61, would be the final of these seven elements to be discovered, in 1945.

### MAN-MADE ELEMENTS

When we turn our attention to the elements with atomic numbers greater than 92, we come to the next stage of development of the periodic table, the man-made elements. These elements were each born out of the Nuclear Age, where scientists crashed relatively light nuclei into one other, either in particle accelerators or nuclear bombs, and where careful analysis of the products of these collisions allowed the isolation of new, heavier nuclei. Pioneers of this work include Americans Al Ghiorso (1915–2010) and Glenn Seaborg (1912–1999), and Russians Georgy Flyorov (1913–1990) and Yuri Oganessian (b. 1933). Seaborg and Oganessian are distinguished in at least one way from all other scientists that ever contributed to the development of the periodic table, since, depending on exactly how one defines naming, they are the only two people to have had elements named after them while they were still alive—seaborgium was named in 1997 when Seaborg was eighty-five years old, and oganesson was named in 2016 when Oganessian was eighty-three.

Alkali Metals
Alkaline Earth Metals
Transition Metals
Post-transition Metals
Metalloids
Lanthanoids
Actinoids
Non-metals
Halogens
Noble Gases
Other

| 1 | 2 | 3 | 4 | 5 | 6 | 7 | 8 | 9 | 10 | 11 | 12 | 13 | 14 | 15 | 16 | 17 | 18 |
|---|---|---|---|---|---|---|---|---|---|---|---|---|---|---|---|---|---|
| 1 H | | | | | | | | | | | | | | | | | 2 He |
| 3 Li | 4 Be | | | | | | | | | | | 5 B | 6 C | 7 N | 8 O | 9 F | 10 Ne |
| 11 Na | 12 Mg | | | | | | | | | | | 13 Al | 14 Si | 15 P | 16 S | 17 Cl | 18 Ar |
| 19 K | 20 Ca | 21 Sc | 22 Ti | 23 V | 24 Cr | 25 Mn | 26 Fe | 27 Co | 28 Ni | 29 Cu | 30 Zn | 31 Ga | 32 Ge | 33 As | 34 Se | 35 Br | 36 Kr |
| 37 Rb | 38 Sr | 39 Y | 40 Zr | 41 Nb | 42 Mo | 43 Tc | 44 Ru | 45 Rh | 46 Pd | 47 Ag | 48 Cd | 49 In | 50 Sn | 51 Sb | 52 Te | 53 I | 54 Xe |
| 55 Cs | 56 Ba | 57-71 | 72 Hf | 73 Ta | 74 W | 75 Re | 76 Os | 77 Ir | 78 Pt | 79 Au | 80 Hg | 81 Tl | 82 Pb | 83 Bi | 84 Po | 85 At | 86 Rn |
| 87 Fr | 88 Ra | 89-103 | 104 Rf | 105 Db | 106 Sg | 107 Bh | 108 Hs | 109 Mt | 110 Ds | 111 Rg | 112 Cn | 113 Nh | 114 Fl | 115 Mc | 116 Lv | 117 Ts | 118 Og |
| | | | 57 La | 58 Ce | 59 Pr | 60 Nd | 61 Pm | 62 Sm | 63 Eu | 64 Gd | 65 Tb | 66 Dy | 67 Ho | 68 Er | 69 Tm | 70 Yb | 71 Lu |
| | | | 89 Ac | 90 Th | 91 Pa | 92 U | 93 Np | 94 Pu | 95 Am | 96 Cm | 97 Bk | 98 Cf | 99 Es | 100 Fm | 101 Md | 102 No | 103 Lr |

▲ The modern IUPAC periodic table, consisting of 118 chemical elements, including the recent formal additions of elements with atomic numbers 113, 115, 117, and 118, which complete the seventh period.

## PERIODS AND "FAMILIES"

The periodic table does far more than simply gather the elements into a single, handy reference guide; it also organizes them. The specific arrangement of the elements is incredibly important, since the rows (known as periods) and columns (known as groups) are not randomly generated—rather, each horizontal and vertical relationship hides a deeper, more profound meaning of chemical similarity and subtle difference. Meaningful chemistry without reference to the periodic table is close to impossible. The modern periodic table now boasts 118 officially recognized and named elements, with the first seven periods complete. Each space on the table shows the element's atomic number, its symbol, and its average atomic weight. Elements in the same group tend to show many similar physical and chemical properties, and as such are sometimes called "families." Elements in any given period show gradual changes as the periodic table is traversed, with similarities often seen between adjacent elements, but with significant differences being apparent once one has crossed the whole table. These general similarities are not without their own anomalies and interruptions, of course, but no matter, the periodic table still provides both chemistry neophytes and seasoned chemical geniuses with profoundly important information about the ragtag collection of unique substances that we call the elements.

This book aims to give a broad overview of the diversity of the elements for a general audience, and as such it is not overly technical. I have opted for a flexible approach that allows for groups to be represented (e.g., the halogens and noble gases), alongside larger, less well-defined collections (e.g., metalloids and post-transition metals). In this way, it is entirely acceptable to highlight uniqueness, and in fact the first element in this book is a good example of that. Element number 1, hydrogen, is a tricky element to pin down in many ways, and it is with it that we start out on our elemental journey.

# Hydrogen

| Chemical symbol | H |
|---|---|
| Atomic number | 1 |
| Atomic mass | 1.008 |
| Boiling point | -423.182°F (-252.879°C) |
| Melting point | -434.49°F (-259.16°C) |
| Electron configuration | 1 |

**Hydrogen is the most abundant element in the universe—it makes up 88 percent of all the atoms present, with helium coming in a distant second at 11 percent. More often than not you'll find it placed above lithium in the periodic table, at the head of group 1 in period 1. Indeed, the official IUPAC periodic table assigns hydrogen that position, but as a gas and not a highly reactive metal, its placement is certainly open to debate.**

## AN EXCEPTIONAL ELEMENT

Hydrogen atoms only contain a single electron. This makes hydrogen a tricky element to categorize. Should the superlight, colorless gas sit with the silver-colored, solid metals of group 1 that also have just one valence electron? Probably not. What about treating hydrogen like a group 17 element, as these, like hydrogen, only require one more electron to complete their valence shell? Either way, there are physical and chemical contradictions, so element number 1 is often treated as an exceptional entity.

## WATER-FORMING

Discovered by Henry Cavendish (1731–1810) in 1766, hydrogen was tantalizingly close to being discovered long before the English eccentric managed to correctly identify it as a unique substance. A number of prominent chemists had described flammable "airs" that were almost certainly hydrogen, notably Robert Boyle (1627–1691). However, partly because real chemical analysis was in its infancy at the time, and partly because one needs a little luck to discover an element, hydrogen went formally undiscovered for decades prior to Cavendish's definitive experiments. He christened element number 1 by demonstrating that when hydrogen burns it forms water—hence "hydrogen," from the Greek *hydro* and *genes*, meaning "water" and "forming," respectively.

The flammability that Cavendish demonstrated has been both a blessing and a curse in the history of hydrogen. The *Hindenburg* disaster was one of its spectacular failures, when, in 1937, a hydrogen-filled airship exploded in a ball of fire in the New Jersey night sky, killing thirty-six people. Hydrogen was used in early airships as the "lifting gas" since its density is approximately only $1/14$ that of air, and its presence therefore creates the buoyancy needed for the aircraft to rise. On the plus side, colorless and odorless $H_2$ gas has enormous promise as an alternative energy source, since when it burns, it only produces harmless water, making it a zero-emission fuel.

## BOTH WATER AND ACID

As the chief component of the sun and stars, and as two-thirds of the atoms that make up a water molecule, hydrogen is vital to earth's very survival. However, it is in the form of its ion, $H^+$, that hydrogen has a huge impact on everyday life. When $H^+$ ions combine with water molecules to form $H_3O^+$ (hydronium) ions, they produce an acidic solution. These substances, and their chemical opposites, bases, are crucially important both in industrial settings and in our homes. A compound such as sulfuric acid, for example, is used extensively in the manufacture of many chemicals, whereas other acids and bases, like lemon juice, battery acid, vinegar, cleaning products, and over-the-counter medicines, play an important role in everyday life.

◂ Atomic hydrogen is the most abundant element in the universe, present in space in stars, including our own sun.

▸ One of the best known and most spectacular examples of the flammability of hydrogen is encapsulated in the *Hindenburg* (a German passenger airship) disaster of 1937.

# ALKALI METALS

**The six elements that comprise the first column (group 1) on the far left of the periodic table are collectively known as the alkali metals. Lithium, sodium, potassium, rubidium, cesium, and francium are sometimes joined by hydrogen at the head of the group, but as we have seen, hydrogen is a unique element, and it is considered by some as being placed there purely for convenience, rather than belonging to the group with any conviction.**

On the following pages:

**Li** Lithium
**Na** Sodium
**K** Potassium
**Rb** Rubidium
**Cs** Cesium
**Fr** Francium

| 1 | 2 | 3 | 4 | 5 | 6 | 7 | 8 | 9 | 10 | 11 | 12 | 13 | 14 | 15 | 16 | 17 | 18 |
|---|---|---|---|---|---|---|---|---|---|---|---|---|---|---|---|---|---|
| 1 H | | | | | | | | | | | | | | | | | 2 He |
| **3 Li** | 4 Be | | | | | | | | | | | 5 B | 6 C | 7 N | 8 O | 9 F | 10 Ne |
| **11 Na** | 12 Mg | | | | | | | | | | | 13 Al | 14 Si | 15 P | 16 S | 17 Cl | 18 Ar |
| **19 K** | 20 Ca | 21 Sc | 22 Ti | 23 V | 24 Cr | 25 Mn | 26 Fe | 27 Co | 28 Ni | 29 Cu | 30 Zn | 31 Ga | 32 Ge | 33 As | 34 Se | 35 Br | 36 Kr |
| **37 Rb** | 38 Sr | 39 Y | 40 Zr | 41 Nb | 42 Mo | 43 Tc | 44 Ru | 45 Rh | 46 Pd | 47 Ag | 48 Cd | 49 In | 50 Sn | 51 Sb | 52 Te | 53 I | 54 Xe |
| **55 Cs** | 56 Ba | 57-71 | 72 Hf | 73 Ta | 74 W | 75 Re | 76 Os | 77 Ir | 78 Pt | 79 Au | 80 Hg | 81 Tl | 82 Pb | 83 Bi | 84 Po | 85 At | 86 Rn |
| **87 Fr** | 88 Ra | 89-103 | 104 Rf | 105 Db | 106 Sg | 107 Bh | 108 Hs | 109 Mt | 110 Ds | 111 Rg | 112 Cn | 113 Nh | 114 Fl | 115 Mc | 116 Lv | 117 Ts | 118 Og |
| | | | 57 La | 58 Ce | 59 Pr | 60 Nd | 61 Pm | 62 Sm | 63 Eu | 64 Gd | 65 Tb | 66 Dy | 67 Ho | 68 Er | 69 Tm | 70 Yb | 71 Lu |
| | | | 89 Ac | 90 Th | 91 Pa | 92 U | 93 Np | 94 Pu | 95 Am | 96 Cm | 97 Bk | 98 Cf | 99 Es | 100 Fm | 101 Md | 102 No | 103 Lr |

## VIOLENT REACTIONS

The alkali metals are noted for their reactivity, which is due to the ease with which they lose their single valence electron, forming the more stable +1 ion in the process. Their reactivity is extreme when compared to other metals. For example, they will react immediately with oxygen when exposed to air, and react violently when they come into contact with water.

The reaction with oxygen can be seen when these soft metals are cut with a knife. One will see an unsullied, shiny surface at the point of the cut, because the internal atoms of the metal have been protected from oxygen. However, once exposed, the bright silvery color will dull within seconds as an oxide layer is built up on the surface of the exposed metal.

Reactions of the group 1 metals with water are equally visible. As particularly light metals, small pieces of them will float on the surface of water and appear to dance around as hydrogen gas is liberated from the water. These reactions are exothermic (ones that release energy), and often the energy released is sufficient to ignite the hydrogen gas and produce what looks like a "floating fire" on the surface of the water—spectacular!

## BODY MATTERS

In their compounds, sodium and potassium in particular have crucial roles in human biology in determining electrical impulses and movement of water in the body. Lithium compounds have proved to be vital medicines in the field of mental illness, being used as antidepressants and mood modifiers.

**The group 1 metals are typified by their reactivity, which increases down the group. They react with water to produce gaseous hydrogen that is ignited in the exothermic process.**

# Lithium

| Chemical symbol | Li |
|---|---|
| Atomic number | 3 |
| Atomic mass | 6.94 |
| Boiling point | 2,448°F (1342°C) |
| Melting point | 356.97°F (180.5°C) |
| Electron configuration | 2.1 |

**The lightest of all of the group 1 elements, lithium exhibits many typical properties of the alkali metal family: It is light, soft, and reactive. If somebody is said to be "on lithium," it generally means that they are receiving a lithium-based medication prescribed for a mental disorder.**

The Swedish chemist Johan August Arfvedson (1792-1841) eventually discovered lithium in 1817. "Eventually," since this was not before a number of earlier chemists had come close to identifying a new element contained within certain rocks. Lithium is found in a number of naturally occurring minerals, notably spodumene and petalite. Several early chemists had suspicions of there being a new element present in these minerals, but it took an analysis of petalite by Arfvedson to confirm it. Lithium's proliferation in such minerals is reflected in its name, which comes from the Greek *lithos*, meaning "stone."

## THE MANIC ELEMENT

It took almost another century and a half before the Australian psychiatrist John Cade (1912–1980) first championed lithium's use as an antidepressant in 1949. His initial experiments with rodents suggested that lithium carbonate could be used as an effective mood stabilizer. Within about twenty years, the use of the carbonate had exploded as a treatment for bipolar disorder in humans, with generally excellent results. Lithium carbonate is still a popular medicine today.

The Swedish chemist J. A. Arfvedson discovered lithium in 1817, during an analysis of a sample of the naturally occurring mineral petalite.

▲ Aluminum–lithium alloy tanks contain liquid oxygen and rocket-grade kerosene, the propellants for the *Falcon 9* spacecraft that has been designed and manufactured by the private U.S. company SpaceX.

◀ Lithium is used in many types of batteries, including these button cells, which are often used to power small devices such as watches.

## LIGHTWEIGHT HIGH-FLIER

As the lightest of all the known metals, lithium has found a major use in metal alloys. Its use here is usually directly related to its light weight and the subsequent reduction in mass that it offers. There are obvious advantages in the aerospace industry, and aluminum–lithium alloys form part of many commercial aircraft. They are also used in the *Falcon 9* space rocket made by SpaceX, the ambitious commercial space-travel company founded by the entrepreneur Elon Musk (b. 1971).

In consumer products, lithium batteries have proved extremely popular. Here lithium acts as the anode (the negative pole in a cell), readily releasing electrons in a typical group 1 manner. Lithium batteries can also exhibit very long lives when compared to other cells, and the combination of lightness and long life often means that despite their relatively expensive nature, they are preferred. Their long life is also utilized in surgically implanted medical devices such as pacemakers, where a need for frequent replacement would be extremely inconvenient.

# Sodium

| | |
|---|---|
| **Chemical symbol** | Na |
| **Atomic number** | 11 |
| **Atomic mass** | 22.99 |
| **Boiling point** | 1,621.292°F (882.94°C) |
| **Melting point** | 208.029°F (97.794°C) |
| **Electron configuration** | 2.8.1 |

**A ubiquitous element when considering compounds, sodium is paradoxically elusive when it comes to its pure, metallic state. As a typical member of group 1, sodium is keen to lose its singular outer electron to achieve a more energetically stable electronic structure. The propensity to achieve this more stable state is so strong that sodium will react with almost anything that it comes into contact with.**

## KEY FIGURE

**HUMPHRY DAVY** 1778–1829

Humphry Davy was a British chemist and a pioneer in the field of electrochemistry—the study of the relationship between electrical and chemical phenomena. Davy discovered sodium in 1807. By passing electricity through sodium hydroxide, he managed to separate the element sodium from the compound, in a process known as electrolysis. Davy discovered potassium in the same way and in the same year, and isolated calcium, strontium, barium, and magnesium the following year via a similar method.

*"[The substance produced] from soda, which was fluid in the degree of heat of the alkali during its formation, became solid on cooling, and appeared having the lustre of silver."*

*— excerpt from Humphry Davy's original paper of 1807, detailing his isolation of sodium*

### EXOTHERMIC REACTION

Sodium's tendency to react is so strong that the soft, silvery, putty-like metal, which can be cut with a knife, is normally stored under oil to prevent even its reaction with air. Its reaction with water is a rapid one that produces the flammable gas hydrogen. This reaction generates heat, and if a large enough piece of sodium is used then the energy produced can cause the hydrogen gas to catch fire.

### ION REGULATOR

Whenever it does react, sodium metal loses one negative electron and forms its positive ion, $Na^{+}$. In this state, sodium finds relative stability, and it will remain benign in common compounds such as sodium chloride (common salt). It is in its ionic state that sodium operates in the human body. As an essential element, sodium ions help regulate the movement of water across the membranes of human cells, and as such are an integral part of many biological functions, including that of the kidneys.

Sodium reacts vigorously with water. Here, the energy released in the exothermic reaction is sufficient to cause an explosion.

The familiar yellow glow of old-fashioned street lighting is due to the sodium vapor in the lamps generating light of a wavelength equal to 589 nanometers, which is in the yellow part of the visible spectrum.

## LIGHTING THE WAY

A far more visible application of sodium is its use in street lighting. Small amounts of solid sodium metal are encased in tubes along with neon and tiny percentages of group 18 elements. As the sodium is gradually heated in the lamp, it vaporizes, and as it does, the movement of electrons within the sodium atoms creates the characteristic yellow light. The wavelength of the light produced in this way is found to be effective in outdoor applications, particularly when fog is present.

# Potassium

| Chemical symbol | K |
|---|---|
| Atomic number | 19 |
| Atomic mass | 39.098 |
| Boiling point | 1,398°F (759°C) |
| Melting point | 146.3°F (63.5°C) |
| Electron configuration | 2.8.8.1 |

**Culturally, potassium is often associated with bananas, since they contain a large amount of the element. There are, however, plenty of other foods that contain larger amounts of potassium, including potatoes, spinach, and some beans. It is one of life's essential elements and crucial to our well-being.**

As a soft, shiny metal, potassium resembles sodium both in appearance and in many of its properties. Another quintessential group 1 element, it will react instantaneously with oxygen and violently with water. Potassium and sodium were essentially "born together" when Humphry Davy perfected the art of electrolysis. This means that sodium and potassium are essentially "twins," but their simultaneous discovery illustrates a far more important point about the nature of the periodic table. When elements are in the same group, their properties are often very similar. When they are adjacent, that similarity can be amplified, so their appearance together is not a coincidence—it occurs because they are so alike. Davy named his newly found element after the substance that he had extracted it from, and that he knew as "potash." The particular potash that he used, we now know as potassium hydroxide. Potash also holds the key to potassium's chemical symbol, K, which always seems somewhat exotic, but is easily explained when one knows that the Latin for potash is *kalium*.

### LIFE AND DEATH

The similarity between elements 11 and 19 is also evident in the human body, where potassium—in its ionic form, $K^+$—plays a similar role to sodium. The regulation of the electrical signals that determine nerve function is just one way in which potassium is an essential element for life. Ironically, potassium continues to be a component of one of the chemicals used to execute humans by lethal injection in those countries where capital punishment still exists. As the third of three drugs administered, potassium chloride is injected to stop the heart after the prisoner has been sedated and paralyzed by other substances.

### BACK TO LIFE

Potassium, again in its ionic form, $K^+$, controls many of the growth processes in plants. Just as the element helps to regulate water flow in humans, a similar role is found in plants, where potassium is central to the process of photosynthesis that converts sunlight into energy. Because of this, potassium is used to improve the general physical quality of vegetation. A huge percentage of the potassium that is produced in the modern world is used to make various potassium-based fertilizers, such as potassium nitrate and potassium sulfate.

◂ Potassium will react violently with water to yield a solution of potassium hydroxide and hydrogen gas. The reaction releases sufficient energy for the hydrogen to ignite.

Potash evaporation ponds, such as this one at a mine in Utah, are used to isolate potassium-containing potash. Sunlight evaporates the water, leaving behind crystallized potassium salts.

# Rubidium

**Chemical symbol**
Rb

**Atomic number**
37

**Atomic mass**
85.468

**Boiling point**
1,270°F (688°C)

**Melting point**
102.74°F (39.3°C)

**Electron configuration**
2.8.18.8.1

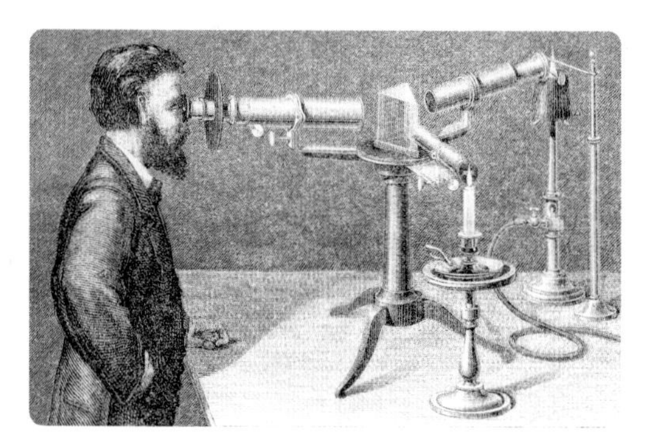

Gustav R. Kirchhoff using an early three-arm spectroscope to identify chemical elements by the characteristic spectrum of radiation they emit when heated.

**Rubidium is often found alongside potassium in naturally occurring minerals in the earth's crust, but, like potassium, because of its reactivity, rubidium is never found as the free metal. Despite its name—from the Latin for deepest red, *rubidius*—rubidium isn't actually red; rather, it is named after the bright red lines observed in its visible spectrum.**

As we make our way down group 1 to the fifth period, the metals are starting to get a little menacing. As we have seen, the group 1 elements are characterized by their reactivity, and the tendency to react increases as the group is descended. Not only does rubidium react violently with water, it has a habit of spontaneously igniting in air. As noted above, rubidium's name is linked to its discovery, in 1861, by Robert Bunsen (1811–1899) (inventor of the Bunsen burner), and his collaborator, Gustav R. Kirchhoff (1824–1887). Along with cesium, rubidium was one of the first two elements discovered using a spectroscope. The spectroscope allowed light from a mineral source to be analyzed, and from the unique spectral lines that were observed, new elements could be identified.

### *NEARLY* ANOTHER LIQUID METAL

The melting point of rubidium is only 103°F (39°C), so holding a piece of the metal in your hand would be sufficient to turn the silvery-white mass into a liquid. Of course, doing so would be foolish, since the metal would react with the moisture on your hand, and you would have a nasty problem involving the production of flammable hydrogen gas!

### NEW HORIZONS

Rubidium gas has been used in the formation of a relatively new state of matter called a Bose–Einstein condensate (BEC). In BECs, gaseous atoms are condensed to produce particles held at temperatures slightly above absolute zero (0 K). In this state, many quantum phenomena become apparent, and advanced studies at the atomic level are possible.

The first ever BEC was produced in 1995, when researchers in Colorado achieved the state by taking approximately 2,000 rubidium-87 atoms and cooling them to 170 nanokelvin, i.e., 0.000000170 K. Soon after, another group at the Massachusetts Institute of Technology used atoms of fellow group 1 element sodium to make a different BEC. The work of the two groups led to the award of the Nobel Prize in Physics in 2001.

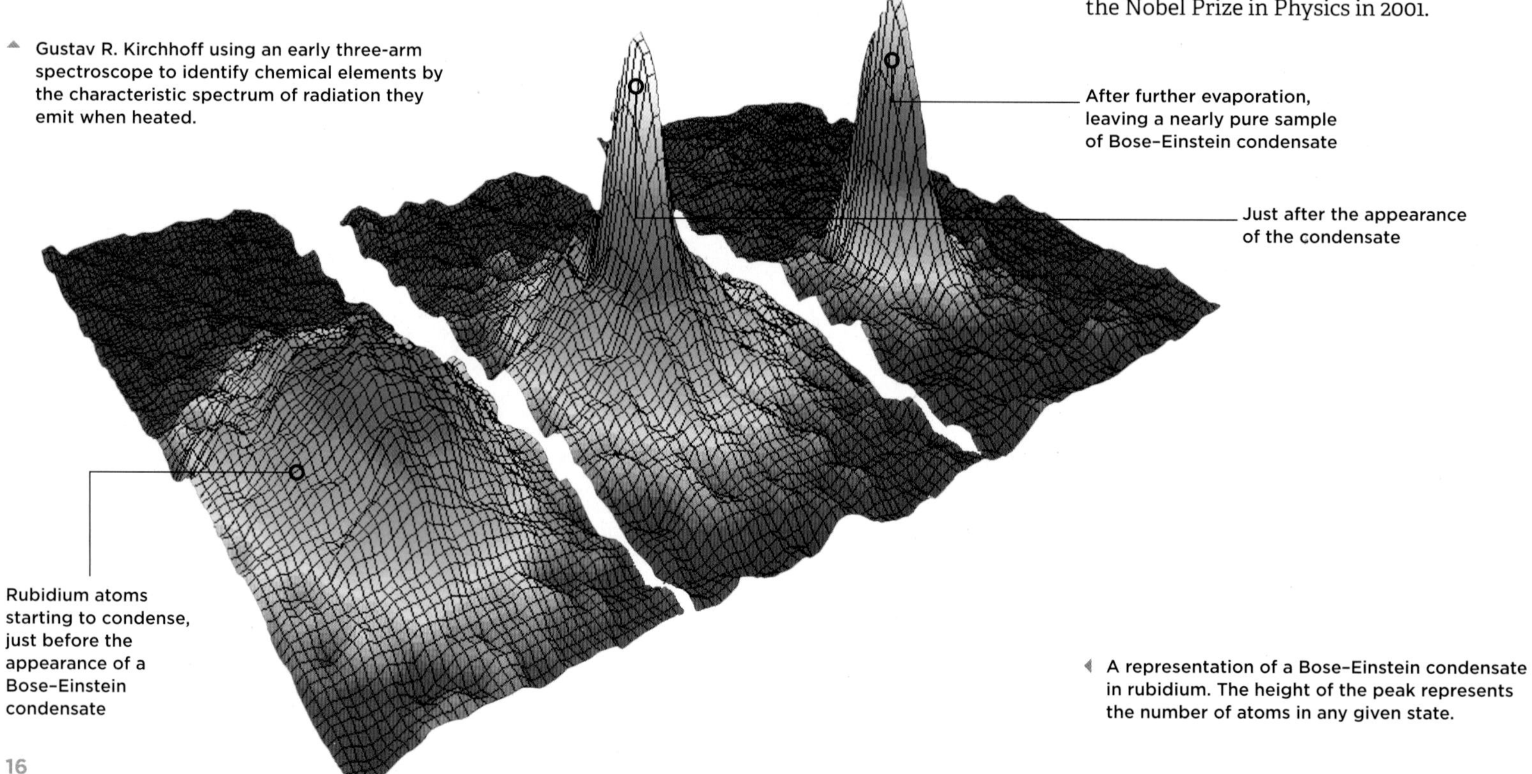


A representation of a Bose-Einstein condensate in rubidium. The height of the peak represents the number of atoms in any given state.

# Cesium

| Chemical symbol | Cs |
|---|---|
| Atomic number | 55 |
| Atomic mass | 132.91 |
| Boiling point | 1,240°F (671°C) |
| Melting point | 83.3°F (28.5°C) |
| Electron configuration | 2.8.18.18.8.1 |

**Cesium is the alkali metal that is most likely to exist as a liquid under ambient conditions. The silvery-gold, soft metal melts at 83°F (28.4°C), just a few degrees above room temperature. Cesium has an elevated ability to lose its negative outer electron in chemical reactions, since the electron is held farther from the positive nucleus than in any of the elements above it in group 1, and can thus be ripped away very easily.**

Cesium was the first element found by Bunsen and Kirchhoff via their spectroscope method. Using samples of mineral water as the source, the two chemists painstakingly removed known compounds from the water until they had concentrated liquor. They saw previously unobserved lines, and put this down to an undiscovered element present. The bright blue lines in cesium's emission spectrum inspired the element's name, since the Latin *caesius* means "sky blue."

## ATOMIC TIME

Cesium's spectral lines are linked to arguably its most important application: in atomic clocks. As with krypton and the meter, the movement of electrons within atoms is such a reproducible event that it can be used as a standard by which SI (International System) units can be defined. In the case of cesium, it is the SI unit of time, the second, that is defined in terms of electron movement. The electronic transition event in the cesium isotope is reliable enough to produce atomic clocks accurate to within one second every several million years.

▲ Jack Parry (left) and Louis Essen (right) designed and built the world's first cesium atomic clock at the UK National Physical Laboratory in 1955.

# Francium

| Chemical symbol | Fr |
|---|---|
| Atomic number | 87 |
| Atomic mass | 223 (longest living isotope) |
| Boiling point | 1,251°F (677°C) |
| Melting point | 81°F (27°C) |
| Electron configuration | 2.8.18.32.18.8.1 |

**Since the reactivity of the alkali metals increases as one descends the group, it is not surprising that francium enjoys a somewhat mythical status as a highly dangerous element. In fact, there is really nothing to worry about. It is not that the predicted ire of francium is inaccurate, but simply that there is estimated to be only a few grams of francium in the earth's crust at any given moment.**

## FALSE STARTS

Francium is another of the seven missing elements highlighted following Moseley's ordering of the periodic table via atomic number. It is a wonder francium was ever discovered at all, given the scarcity of it, and the incredibly short half-lives exhibited by its isotopes. Francium was commonly known as eka-cesium at the outset of the search, and over the years an unusually large number of false claims were made of its discovery.

Eventually, in 1939, a French researcher solved the mystery of the missing element. Marguerite Perey (1909–1975) was studying the radioactive nature of actinium, and noticed some unusual beta emission coming from her samples. She deduced that this radioactivity was not from the actinium itself, but from another source. That source turned out to be the elusive francium.

▲ French physicist Marguerite Perey discovered francium in 1939 and named the element after her native country. Perey had been a student of Marie Curie.

# ALKALINE EARTH METALS

**The group 2 elements share many similarities with their group 1 neighbors to the left, inasmuch as they are not found free in nature and are reactive, but each of the extreme properties found in group 1 is toned down a little for group 2. A little harder, a little more dense, and a little less reactive than the alkali metals, the alkaline earth metals also have higher melting and boiling points. Compared to metals not in group 1, however, the alkaline earths are fairly soft and quite reactive.**

On the following pages:

**Be** Beryllium
**Mg** Magnesium
**Ca** Calcium
**Sr** Strontium
**Ba** Barium
**Ra** Radium

| 1 | 2 | 3 | 4 | 5 | 6 | 7 | 8 | 9 | 10 | 11 | 12 | 13 | 14 | 15 | 16 | 17 | 18 |
|---|---|---|---|---|---|---|---|---|---|---|---|---|---|---|---|---|---|
| 1 H | | | | | | | | | | | | | | | | | 2 He |
| 3 Li | **4 Be** | | | | | | | | | | | 5 B | 6 C | 7 N | 8 O | 9 F | 10 Ne |
| 11 Na | **12 Mg** | | | | | | | | | | | 13 Al | 14 Si | 15 P | 16 S | 17 Cl | 18 Ar |
| 19 K | **20 Ca** | 21 Sc | 22 Ti | 23 V | 24 Cr | 25 Mn | 26 Fe | 27 Co | 28 Ni | 29 Cu | 30 Zn | 31 Ga | 32 Ge | 33 As | 34 Se | 35 Br | 36 Kr |
| 37 Rb | **38 Sr** | 39 Y | 40 Zr | 41 Nb | 42 Mo | 43 Tc | 44 Ru | 45 Rh | 46 Pd | 47 Ag | 48 Cd | 49 In | 50 Sn | 51 Sb | 52 Te | 53 I | 54 Xe |
| 55 Cs | **56 Ba** | 57-71 | 72 Hf | 73 Ta | 74 W | 75 Re | 76 Os | 77 Ir | 78 Pt | 79 Au | 80 Hg | 81 Tl | 82 Pb | 83 Bi | 84 Po | 85 At | 86 Rn |
| 87 Fr | **88 Ra** | 89-103 | 104 Rf | 105 Db | 106 Sg | 107 Bh | 108 Hs | 109 Mt | 110 Ds | 111 Rg | 112 Cn | 113 Nh | 114 Fl | 115 Mc | 116 Lv | 117 Ts | 118 Og |
| | | | 57 La | 58 Ce | 59 Pr | 60 Nd | 61 Pm | 62 Sm | 63 Eu | 64 Gd | 65 Tb | 66 Dy | 67 Ho | 68 Er | 69 Tm | 70 Yb | 71 Lu |
| | | | 89 Ac | 90 Th | 91 Pa | 92 U | 93 Np | 94 Pu | 95 Am | 96 Cm | 97 Bk | 98 Cf | 99 Es | 100 Fm | 101 Md | 102 No | 103 Lr |

## THE EARTHS

Members of group 2 were first given the name "earths," and there was much conjecture as to whether what ultimately turned out to be oxides were indeed elements at all. Even when it was established that the oxides were in fact compounds that could yield the group 2 elements, the original name stuck, and beryllium, magnesium, calcium, strontium, barium, and radium became the alkaline earths. As with group 1, the alkali/alkaline part of the name comes from the fact that the oxides and hydroxides of the group 1 and group 2 elements are chemical bases, and the fact that an alkali is a base.

## HEALTH AND WELLNESS

Magnesium and calcium are both essential elements to humans and other life-forms. As the metal at the center of chlorophyll—the green pigment that allows plants to convert light to energy—magnesium is vital; as a crucial component of bones and teeth, calcium is central to skeletal health. Radium, on the other hand, is a ferociously radioactive element that has done much damage to health over the years. Magnesium is a dangerously flammable element (unusual for a metal) that burns with a bright, white light. The salts of strontium, calcium, and barium are used in fireworks, since they provide stunning colors when heated.

Salts of the alkaline earth metals strontium, calcium, and barium are used in fireworks since they produce vivid colors.

# Beryllium

**Chemical symbol**
Be

**Atomic number**
4

**Atomic mass**
9.0122

**Boiling point**
4,474°F (2,468°C)

**Melting point**
2,349°F (1,287°C)

**Electron configuration**
2.2

**The lightest of the group 2 elements, beryllium gets its name from the mineral beryl. For over 150 years element number 4 was named glucinium from "glucina," the word used to describe the oxide (or earth) of beryllium. That word in turn is derived from the Greek *glukus*, meaning "sweet." Apparently, beryllium compounds are sweet to the taste.**

But tasting as a method of chemical analysis is ill advised, especially when handling beryllium. Long-term exposure to the element and its compounds can lead to the lung disease berylliosis, whose symtoms include shortness of breath, coughing, and chest pain. Workers employed in the fluorescent light industry, which used beryllium until the practice was discontinued in 1949, were vulnerable to the disease. In addition to berylliosis, exposure to beryllium can cause cancer, long-term poisoning, and skin problems. On the flip side, beryllium is certainly an attractive element. Both in its steely-gray metallic form and in the brilliant blues and greens of the aquamarine and emerald gemstones that contain the oxide, element number 4 has its moments.

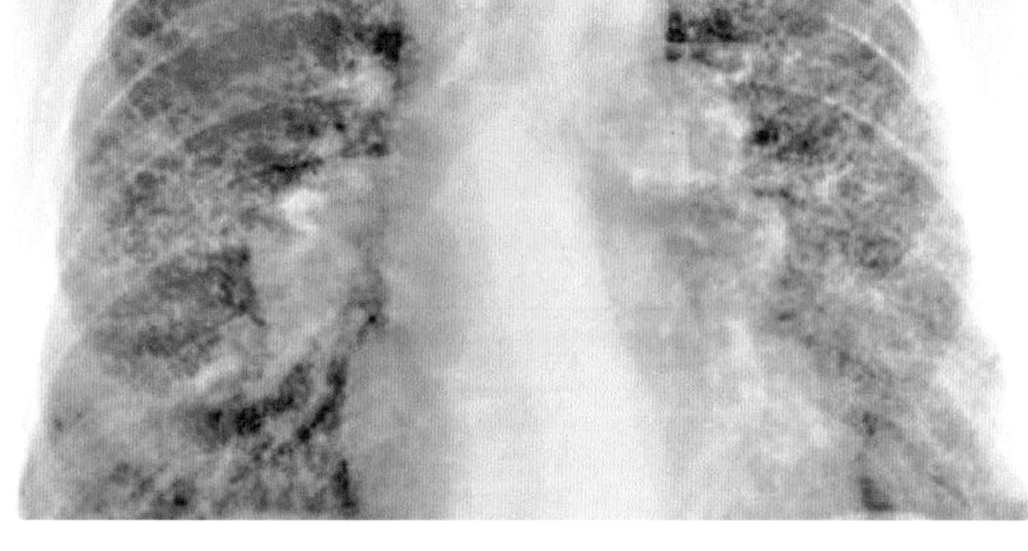

The occupational disease berylliosis is a chronic lung condition that is caused by exposure to beryllium and its compounds.

## NEUTRON DISCOVERY

Beryllium was at the heart of the discovery of the neutron in 1932. As part of the process that James Chadwick used in his experiments that led to the discovery of the neutral subatomic particle, a beryllium nucleus was bombarded with alpha particles, and a carbon atom and a neutron were the result:

$${}^{9}_{4}\mathrm{Be} + {}^{4}_{2}\mathrm{He} \rightarrow {}^{12}_{6}\mathrm{C} + {}^{1}_{0}\mathrm{n}$$

In more modern chemistry, beryllium has a couple of important applications. With copper, in an alloy containing only about 2 percent beryllium, it produces a much harder, more resilient version of pure copper, while maintaining the excellent conductivity of the heavier metal. In a more sinister role, it is used in the manufacture of nuclear weaponry. The beryllium is used to encase the fissile material where it acts as a neutron reflector, trapping the neutrons that cause the chain reaction. In turn, the element also acts as a tamper, momentarily containing the initial chain reaction, making its ultimate release even more devastating.

Aquamarine is a precious variety of beryl, a mineral ore of beryllium.

# Magnesium

**Chemical symbol**
Mg

**Atomic number**
12

**Atomic mass**
24.305

**Boiling point**
1,994°F (1,090°C)

**Melting point**
1,202°F (650°C)

**Electron configuration**
2.8.2

## KEY FIGURE

**JOSEPH BLACK** 1728–1799

Scottish chemist and physician Joseph Black "discovered" magnesium by distinguishing between magnesia and lime (the oxides of magnesium and calcium, respectively). He also identified carbon dioxide (then known as "fixed air") as a unique gas and investigated its properties, and he pioneered work on latent heat of fusion, latent heat of vaporization, and specific heat capacity—all important concepts in physical chemistry.

**As an important structural metal, and via its use in alloys, element number 12 combines strength with a relatively low density to make it an attractive metal where such a combination is desirable—in the aerospace industry, for example. Discovered in 1755 by the Scottish chemist Joseph Black, it was another of the elements isolated via electrolysis by Humphry Davy in 1808.**

In its silvery-gray metallic form, magnesium is highly reactive. It is also extremely flammable when in its powder form. Perhaps the most infamous magnesium fire of all time occurred during the 1955 Le Mans 24-hour race, when a car crashed into the crowd, killing its driver, Pierre Levegh, and eighty-three spectators. The car's bodywork contained a high magnesium content and rescue workers, unaware of the chemical reaction that would be caused, attempted to extinguish the fire with water. The magnesium reacted with steam to release flammable hydrogen gas, and the fire was greatly intensified.

Magnesium's potentially hazardous nature is contradicted by its absolutely essential role in animal and plant life. As a component of a number of enzyme-based reactions in animals, it helps to regulate energy transfer, muscle action, and a host of other biological functions. The metal also sits at the center of the chlorophyll molecule that plants use in photosynthesis when they convert light to energy.

### SACRIFICIAL PROTECTION

Like zinc, magnesium's reactivity can be put to use in the role of a sacrificial metal. Magnesium's propensity to release electrons more readily than other metals means it can protect those other metals from corrosion. In almost every example of the alloying of magnesium, the strength-to-weight advantages of the group 2 element are exploited. With aluminum, it produces the alloys magnox and magnalium. Magnalium offers enhanced strength without the addition of mass, and is used in aircraft, automobile, and bicycle frame manufacture.

### PERILOUS PRODUCTION

A lot of magnesium is produced via electrolysis, a process that uses electricity. The cost means that recycling the metal is important, and magnesium recycling plants are common. With a large amount of magnesium concentrated in one place, those plants can pose a significant fire hazard, and in two separate incidents in Ohio, in 2003 and 2012, massive infernos occurred at magnesium recycling plants. Magnesium was a crucial component of the flash powder used in early photography. Lighting the explosive mixture of powdered magnesium and the strong oxidizing agent potassium chlorate by hand, with a naked flame, often proved an adventure!

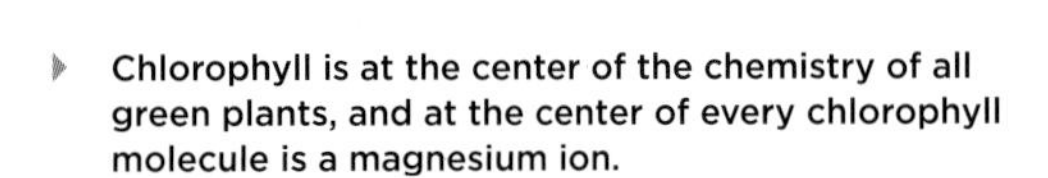

Chlorophyll is at the center of the chemistry of all green plants, and at the center of every chlorophyll molecule is a magnesium ion.

# Calcium

| Chemical symbol | Ca |
|---|---|
| Atomic number | 20 |
| Atomic mass | 40.078 |
| Boiling point | 2,703°F (1,484°C) |
| Melting point | 1,548°F (842°C) |
| Electron configuration | 2.8.8.2 |

**Calcium is far more likely to be encountered in everyday life as one of its compounds than in its elemental state. As an element, it is a silvery metal like its fellows in group 2, but in most of its compounds it appears as a hard, white solid. Its most common forms on earth include gypsum, marble, and lime—calcium sulfate, calcium carbonate, and calcium oxide, respectively.**

Lime was one of Antoine Lavoisier's (1743–94) original elements—more precisely, one of his "earths." It took until 1808, when Humphry Davy isolated calcium via electrolysis, before calcium's status as an element was proven.

#### BEYOND THE SKELETAL SYSTEM

Most people think of calcium as being an essential element for health because of the compounds that it forms in teeth and bones. But calcium is so much more than simply an element of structure in the body. It has a role in transmitting nerve signals via its ionic conductivity, and has an important job in regulating the pH of blood. Compounds of calcium are used in agriculture to regulate acidity in soil. The chief compound used in that role is calcium hydroxide, $Ca(OH)_2$, also known as slaked lime. Slaked lime is derived from quicklime, calcium oxide, by the addition of water. It has been used as a building material for thousands of years, and is still a hugely important chemical in the construction industry (as a component of cement) and in a host of other applications.

The familiar buildup of limescale in household electrical appliances that are exposed to hard water is caused by the presence of calcium ions.

## HARD WATER

Along with the element directly above it, magnesium, calcium imparts a property to water that huge resources are devoted to eradicating. The phenomenon, known simply as "hard water," is caused by the presence of $Ca^{2+}$ (and $Mg^{2+}$) ions. These ions have several problematic effects, including making it more difficult to form lathers with soaps because of the formation of insoluble stearates, or soap "scums," and the buildup of "limescale" on the components of water-heating elements such as those found in electric kettles.

Various calcium compounds can be found in building materials such as concrete.

# Strontium

**Chemical symbol**
Sr

**Atomic number**
38

**Atomic mass**
87.62

**Boiling point**
2,510.6°F (1,377°C)

**Melting point**
1,431°F (777°C)

**Electron configuration**
2.8.18.8.2

**A close relative of calcium, strontium also has a significant historical connection to the element above it, and indeed to element number 56 directly below it. As Humphry Davy continued to romp through the periodic table—isolating metals left, right, and center via electrolysis—strontium, like calcium and barium, was one of his conquests. He first isolated the metal in 1808.**

In terms of chemistry, Scotland has a few claims to fame. Two of the most obvious reside among the alkaline earth metals of group 2. Along with Scotsman Joseph Black, who is credited with the discovery of magnesium, Scotland also yielded strontium, from a mineral first found in the small mining village of Strontian in the West Highlands. A Scottish doctor by the name of Adair Crawford (1748–1795) noticed that the mineral, initially thought to be a barium compound but ultimately found to be strontium carbonate, was distinct from similar barium minerals in terms of its properties. Thus, element number 38 was born.

The "Baker" explosion, part of Operation Crossroads, a nuclear weapon test carried out by the U.S. military at Bikini Atoll, Marshall Islands, in 1946.

## A SINISTER ISOTOPE

In keeping with the other group 2 elements, strontium is another soft, silvery, reactive metal that will quickly form an oxide layer on its surface. It exists as a number of non-radioactive isotopes in nature, but it has one isotope whose reputation has gone before it: strontium-90. Strontium-90 can pose a serious threat to human health. The insidious nature of the isotope lies in its ability to mimic calcium, and thus find its way into bones. Once inside, strontium-90 can do untold damage to the bone, the bone marrow, and the surrounding tissue. Strontium-90 is not a naturally occurring isotope of element number 38, so where does it come from? Like cobalt-60, strontium-90 is a product of nuclear reactions, specifically from the fallout from nuclear testing that took place in the late 1940s and early 1960s in the United States and the Marshall Islands in the Pacific. This testing caused the distribution of strontium-90 over a wide area, and the isotope subsequently made its way into the food chain via plants and animals. In 1959 the Greater St. Louis Citizens' Committee for Nuclear Information initiated the collection of hundreds of thousands of children's baby teeth, and it was quickly established that there had been a significant accumulation of strontium-90 in the children. The outcry surrounding the discovery led to a ban on aboveground nuclear testing.

## FILLING HOLES

In the manufacture of toothpaste for sensitive teeth, strontium acetate, $Sr(CH_3COO)_2$, and strontium chloride, $SrCl_2$, utilize the fact that strontium can act like calcium, to help fill the microscopic holes in teeth.

# Barium

**Chemical symbol**
Ba

**Atomic number**
56

**Atomic mass**
137.33

**Boiling point**
3,353°F (1,845°C)

**Melting point**
1,341°F (727°C)

**Electron configuration**
2.8.18.18.8.2

**Barium's name is derived from the Greek *barys*, meaning "heavy," but that moniker is falsely applied to the element itself. Barium is a relatively light element. Where the heaviness is manifest is in the compounds of barium. Sitting toward the bottom of group 2, barium is a highly reactive element, and as such is not found free in nature.**

One of the most important of those compounds is barium sulfate, $BaSO_4$. Barium sulfate is found in nature as the mineral barite, and is a very dense substance. In oil drilling, barium sulfate is added to the "mud" used around the well and drill bit to help them withstand the pressures created at such depths.

## BARIUM X-RAYS

Most of the tissue that makes up the human gut will allow X-rays to pass straight through, so it is not possible to use the rays as a diagnostic tool for the digestive system in the way that they can be used to image human bones. In steps barium sulfate, in the form of the ominous-sounding "barium enema." Barium sulfate can either be swallowed, or inserted into the gut via the anus. Both methods fill the gastrointestinal tract with the impenetrable barium salt, and if exposed to X-rays, it will produce useful images for medical diagnostics. In other, more soluble salts, such as chloride, bromide, and iodide, barium is a dangerously toxic element that can cause vomiting, diarrhea, and other digestive ailments. If one is going to ingest barium, it's crucial to get the right compound!

## COMPOUNDS MADE IN SITU

Barium's ability to impart pleasing colors to flames is used in fireworks, and the compound most often used is barium chloride. Barium chloride is hygroscopic, meaning that it easily absorbs water from the atmosphere. This makes the compound "wet," which creates a problem when attempting to set fire to fireworks. The solution involves introducing other compounds of barium and chlorine into the firework, and allowing the formation of barium chloride in the gaseous phase once the firework is lit. In that way, the fireworks can be easily ignited and the characteristic green color of barium salts can light up the night sky.

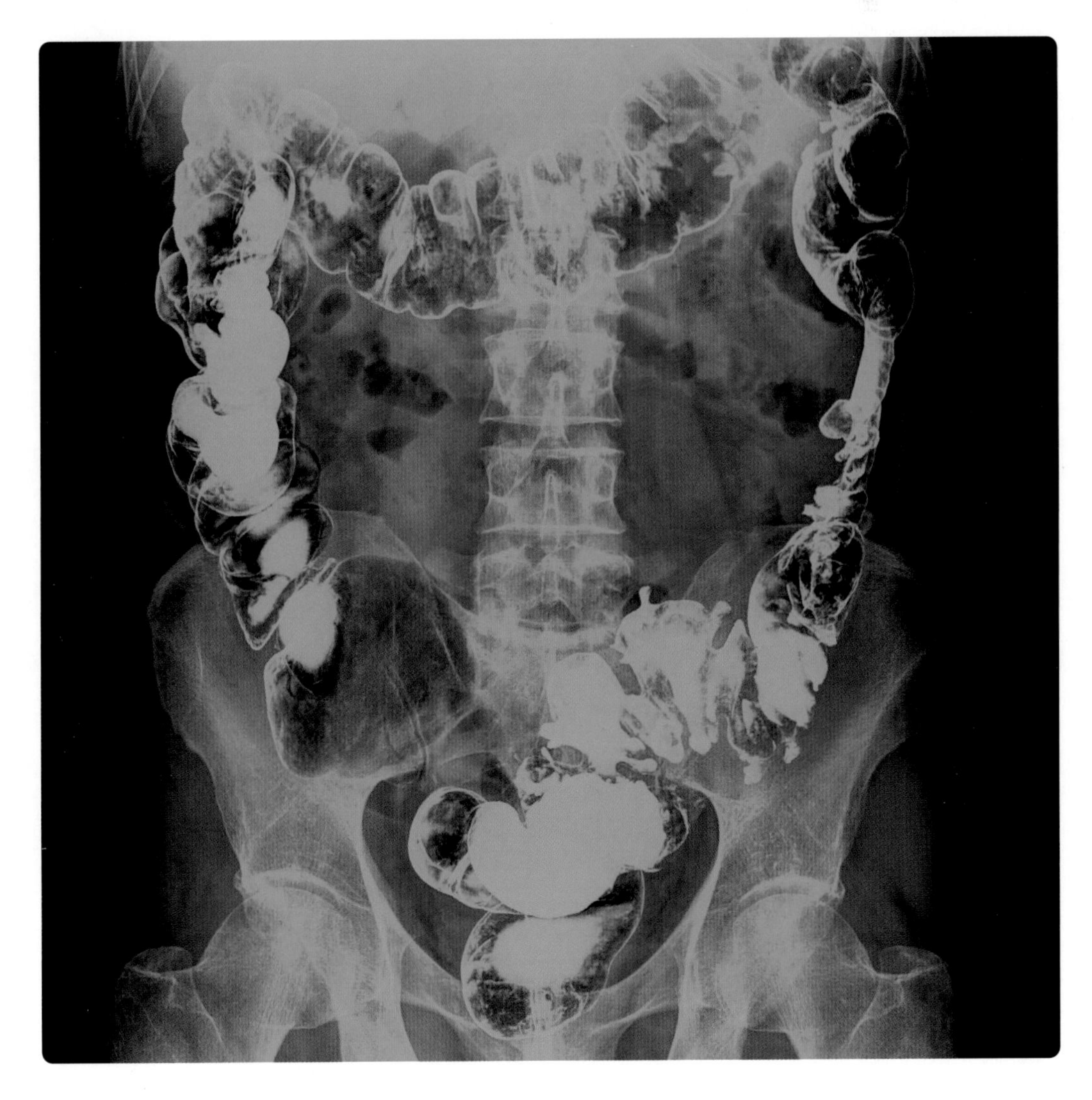

So-called barium meals are given to patients as a diagnostic tool to allow investigation of the stomach and small intestine with X-rays.

# Radium

| | |
|---|---|
| **Chemical symbol** | Ra |
| **Atomic number** | 88 |
| **Atomic mass** | 226 (longest-living isotope) |
| **Boiling point** | 3,159°F (1,737°C) |
| **Melting point** | 1,292°F (700°C) |
| **Electron configuration** | 2.8.18.32.18.8.2 |

**Radium was discovered and named by Marie Curie (1867–1934) and her husband, Pierre (1859–1906), in Paris in 1898. Inextricably linked with radioactivity, radium has a truly extraordinary history of bizarre use in "medicine," being represented at one time or another as a cure-all for many ailments and conditions including, ironically, cancer.**

Like cesium and strontium before it, radium was identified via the unique spectrum that it produced on spectroscopic analysis. It was named for the blue glowing rays of light that the element produces in air, and it is this property that was put to use in one of radium's most notorious applications.

## RADIUM DIALS

The Radium Luminous Material Corporation used radium, in combination with other chemicals, in its luminous paints. In the early part of the twentieth century, these paints were used to produce clock, watch, and dial faces that would glow in the dark. One technique commonly used by the young women employed to apply the paint was to bring the paintbrushes to a fine point by licking them. The prolonged and intense exposure to the radioactive radium in the paint caused terrible radiation sickness, cancers, and many deaths among the workers. The fact that the owners of the company knew about the risks but attempted to cover up the dangers led to huge controversy and court proceedings, with the company eventually settling with the women. However, that was not before the devastating effects of radium poisoning had gripped the lives of so many.

◂ Marie and Pierre Curie discovered radium in 1898 during their pioneering work on radioactivity. They named the new element after the Latin word for "ray."

▴ The Revigator was marketed as a cure for a large number of ailments before its exposure as a health hazard.

The "radium girls" working in a watch factory. The young women contracted radiation poisoning from painting dials with self-luminous paint containing radium.

## THE REVIGATOR

The controlled and more studious use of radium as a treatment for cancer makes it an important element in twenty-first-century medicine, but a hundred years earlier, soon after its discovery, it was used in an almost completely indiscriminate manner. Touted by many as a tonic for general good health, it was marketed as a cure for a large number of ailments via the "Radium Ore Revigator." The Revigator was a ceramic container that was coated on the inside with a whole host of dubious chemicals, among them a uranium ore called carnotite. Traces of radium were present in the ore, and the idea was that water would be stored in the device overnight, and that the resultant radioactive concoction would invigorate the water and promise good health! The advertising told potential customers that the water would be infused with "the lost element of original freshness—radioactivity." In the 1920s and '30s, several hundreds of thousands of these devices were sold before the hazards associated with them were finally brought to light. In a later analysis, it was determined that the arsenic and lead that were also present in the jars were likely to be equally to blame for the sickness that the "medical" device inflicted.

# TRANSITION METALS

The elements in groups 3 through 12 are collectively known as the transition metals. English chemist Charles R. Bury (1890–1968) coined the name in 1921, when he referred to the sequence of metals starting at titanium and passing through to copper as a *transition series*. He was describing the "transition" from eight electrons in the third electron layer to a saturated layer of eighteen.

On the following pages:

**Sc** Scandium
**Ti** Titanium
**V** Vanadium
**Cr** Chromium
**Mn** Manganese
**Fe** Iron
**Co** Cobalt
**Ni** Nickel
**Cu** Copper
**Zn** Zinc
**Y** Yttrium
**Zr** Zirconium
**Nb** Niobium
**Mo** Molybdenum
**Tc** Technetium
**Ru** Ruthenium
**Rh** Rhodium
**Pd** Palladium
**Ag** Silver
**Cd** Cadmium
**Hf** Hafnium
**Ta** Tantalum
**W** Tungsten
**Re** Rhenium
**Os** Osmium
**Ir** Iridium
**Pt** Platinum
**Au** Gold
**Hg** Mercury

| 1 | 2 | 3 | 4 | 5 | 6 | 7 | 8 | 9 | 10 | 11 | 12 | 13 | 14 | 15 | 16 | 17 | 18 |
|---|---|---|---|---|---|---|---|---|---|---|---|---|---|---|---|---|---|
| 1 H | | | | | | | | | | | | | | | | | 2 He |
| 3 Li | 4 Be | | | | | | | | | | | 5 B | 6 C | 7 N | 8 O | 9 F | 10 Ne |
| 11 Na | 12 Mg | | | | | | | | | | | 13 Al | 14 Si | 15 P | 16 S | 17 Cl | 18 Ar |
| 19 K | 20 Ca | 21 Sc | 22 Ti | 23 V | 24 Cr | 25 Mn | 26 Fe | 27 Co | 28 Ni | 29 Cu | 30 Zn | 31 Ga | 32 Ge | 33 As | 34 Se | 35 Br | 36 Kr |
| 37 Rb | 38 Sr | 39 Y | 40 Zr | 41 Nb | 42 Mo | 43 Tc | 44 Ru | 45 Rh | 46 Pd | 47 Ag | 48 Cd | 49 In | 50 Sn | 51 Sb | 52 Te | 53 I | 54 Xe |
| 55 Cs | 56 Ba | 57-71 | 72 Hf | 73 Ta | 74 W | 75 Re | 76 Os | 77 Ir | 78 Pt | 79 Au | 80 Hg | 81 Tl | 82 Pb | 83 Bi | 84 Po | 85 At | 86 Rn |
| 87 Fr | 88 Ra | 89-103 | 104 Rf | 105 Db | 106 Sg | 107 Bh | 108 Hs | 109 Mt | 110 Ds | 111 Rg | 112 Cn | 113 Nh | 114 Fl | 115 Mc | 116 Lv | 117 Ts | 118 Og |
| | | | 57 La | 58 Ce | 59 Pr | 60 Nd | 61 Pm | 62 Sm | 63 Eu | 64 Gd | 65 Tb | 66 Dy | 67 Ho | 68 Er | 69 Tm | 70 Yb | 71 Lu |
| | | | 89 Ac | 90 Th | 91 Pa | 92 U | 93 Np | 94 Pu | 95 Am | 96 Cm | 97 Bk | 98 Cf | 99 Es | 100 Fm | 101 Md | 102 No | 103 Lr |

## CLASSIFICATION CONFUSION

You may think that we are simply talking about elements 21–30 in the fourth period, 39–48 in the fifth period, 72–80 in the sixth period, and 104–112 in the seventh, but of course it is never quite that simple. To make matters more complicated, that same section of the periodic table also goes by another name, the "d-block," which refers to the fact that within that section, d-orbitals are being filled.

One definition of transition metals says that they are elements that exhibit partially filled d-subshells in their ions. This presents a problem for elements such as scandium and zinc that commonly form ions where the d-subshells are completely empty and entirely filled, respectively. So, as with our other "forced" groupings, we need to apply a little latitude rather than fulfilling a rigid requirement.

## CHEMICAL CHARACTERISTICS

So what commonalities can we safely apply to the transition metals that have been chosen for discussion in this chapter? In the simplest terms, these are the elements in the fourth, fifth, and sixth periods, whose final electron enters the 3d, 4d, or 5d sublevel. The similarities in the electronic configurations of these malleable, high-melting-point, lustrous metals lend them a few characteristic properties. Such properties include their ability to exhibit multiple oxidation states, their inclination to form complex ions, and their use as catalysts. As complex ions such as $[Cu(H_2O)_6]^{2+}$ (blue) and $[CuCl_4]^{2-}$ (yellow), transition metals bring color to aqueous solutions.

One highly visible characteristic of transition metals is their ability to form a wide variety of brightly colored salts.

# Scandium

**Chemical symbol**
Sc

**Atomic number**
21

**Atomic mass**
44.956

**Boiling point**
5,136°F (2,836°C)

**Melting point**
2,806°F (1,541°C)

**Electron configuration**
2.8.9.2

**Scandium is an element with a reputation for being a delicate influencer. A soft, silver-colored metal, it is often alloyed with other metals to subtly change their properties, and in the process, to make those other metals more useful.**

Scandium gets its name from the Latin for Scandinavia, *Scandia*, where Lars Fredrik Nilson (1840–1899) first discovered it in 1879. Mendeleev had left a gap in his periodic table for an element that would have an atomic mass of approximately 45, and that he called eka-boron. It turned out that scandium matched the predicted properties of eka-boron very closely, and thus another of the Russian's predicted elements had been found.

## HIGH-PERFORMANCE ALLOYS

Scandium is found in many ores in a number of places on earth, but none of the sources are particularly plentiful. As a result, scandium is not cheap and it means that scandium tends to find few widespread commercial applications. However, one important use of the metal is in an alloy that it makes with aluminum. Even very small amounts of scandium can dramatically increase the hardness of the lightweight metal, making it suitable in applications where pure aluminum might not be viable. This scandium–aluminum alloy has been used extensively in constructing fighter aircraft, and has found use in high-performance sports equipment.

## TRIPLE CHARGE

Much of scandium's chemistry revolves around the +3 oxidation state it achieves when it loses its 4s and 3d electrons. In that +3 state, scandium forms a compound with iodine, $ScI_3$, which has an important application. Since the late 1960s, scandium(III) iodide has been used to create high-intensity lightbulbs that have an interesting property: they mimic natural sunlight very well. This makes them useful in applications such as television and photography, where natural light generally produces better results. Scandium exists on earth as only one isotope, scandium-45, but a synthetic radioactive isotope, scandium-46, has found use as a tracer. By introducing radioactive atoms into a system where fluids move through pipes, the flow can be monitored. This has been utilized in the oil industry, where leaking pipes can be identified by the detection of the radioactive isotope. Although subject to regulation, if isotopes with relatively short half-lives of less than 120 days are used (scandium-46 has a half-life of approximately 84 days), then any risk to health or the environment is minimized.

Scandium's role as a high-performance metal is exemplified by its use in modern jet-fighter aircraft.

# Titanium

| | |
|---|---|
| **Chemical symbol** | Ti |
| **Atomic number** | 22 |
| **Atomic mass** | 47.867 |
| **Boiling point** | 5,949°F (3,287°C) |
| **Melting point** | 3,034°F (1,668°C) |
| **Electron configuration** | 2.8.10.2 |

**Titanium is synonymous with toughness since element number 22 is a material with one of the highest strength-to-weight ratios among the transition metals. The element's strength is, of course, the source of its name, since the metal was christened after the Greek gods known as the Titans, who were revered for their incredible strength.**

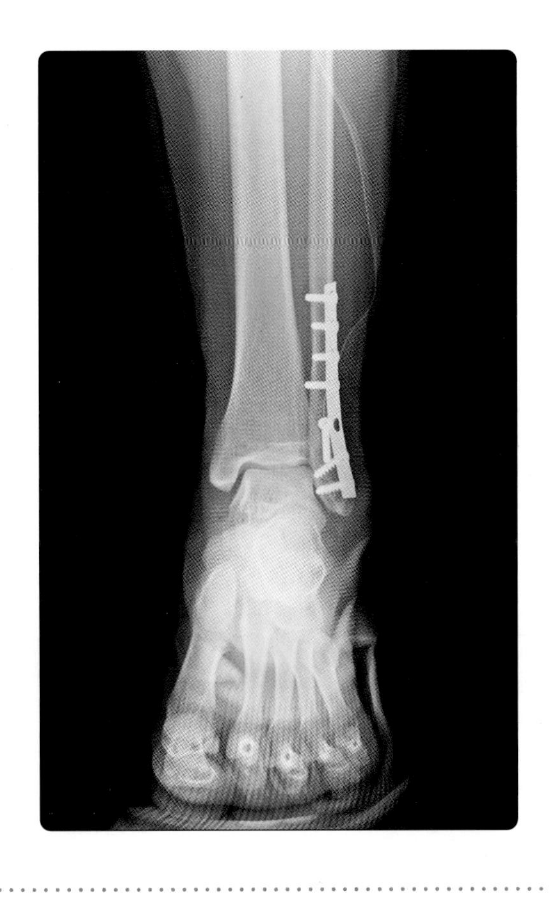

Many people have had bones repaired with titanium plates and screws. The element's corrosion resistance and high strength make it a good choice for surgical repairs.

### STRONG AND LIGHT

As a metal with high strength and relative lightness, titanium is immensely useful, whether in its pure metal form or in alloys. Many parts of jet engines are constructed from titanium and its alloys, and not only for the strength-to-weight advantages: Titanium is also incredibly resistant to corrosion.

### TITANIUM BODY PARTS

Titanium is highly resistant to corrosion because, like aluminum, it rapidly forms an oxide layer on the surface that protects the pure metal underneath from attack. Its ability to resist oxidation means that it is a metal with important medical applications, as it can be implanted in various ways inside the body. Since it is used to produce replacement hip joints, and the screws and plates used to repair badly broken bones, there are a lot of people walking around with titanium inside them.

# Vanadium

| | |
|---|---|
| **Chemical symbol** | V |
| **Atomic number** | 23 |
| **Atomic mass** | 50.942 |
| **Boiling point** | 6,165°F (3,407°C) |
| **Melting point** | 3,470°F (1,910°C) |
| **Electron configuration** | 2.8.11.2 |

**Vanadium is a quintessential example of a transition metal, exhibiting many typical properties: the ability to form multiple oxidation states (charges), its use in catalysts, and its ability to exhibit different colors in its various compounds. As vanadium passes through its oxidation states, spectacular color changes can be observed, from purple, to green to blue, and finally to yellow.**

### COLORFUL HISTORY

Vanadium has gone by a variety of names, several of which related to the vivid colors of its compounds. One example is erythronium, derived from the Greek word for red, *erythros*. As such, it is appropriate that the story of vanadium's discovery is somewhat colorful.

Vanadium was "discovered" twice—first by Andrés Manuel del Río (1764–1849) in 1801, only for him to be convinced that he hadn't actually found a new element, but simply encountered the already known (and equally colorful) chromium. In 1830, Swedish chemist Nils Sefström (1787–1845) "rediscovered" vanadium in some iron compounds and named it after the Norse goddess Vanadis, renowned for her beauty.

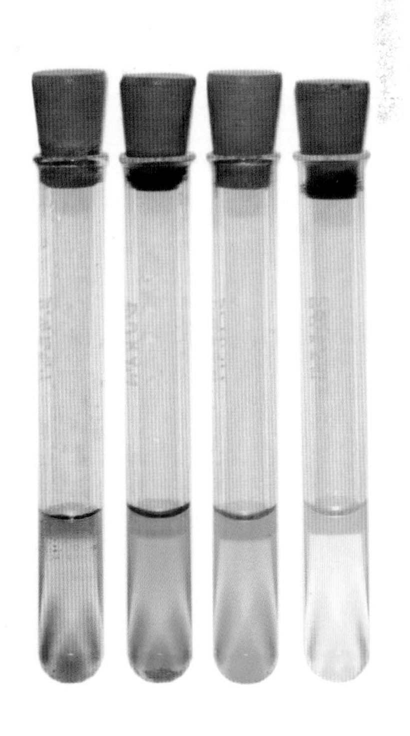

Vanadium is renowned for its multicolored oxidation states, seen here, from left to right, in the +5, +4, +3, and +2 states.

# Chromium

**Chemical symbol**
Cr

**Atomic number**
24

**Atomic mass**
51.996

**Boiling point**
4,840°F (2,671°C)

**Melting point**
3,465°F (1,907°C)

**Electron configuration**
2.8.13.1

**Chromium's very name, taken from the Greek *chroma*, meaning "color," suggests a rich palette of hues. Indeed, element number 24 exhibits one of the quintessential properties of a transition metal: the ability to form colored compounds. But it is as the equally attractive pure metal that chromium shines—quite literally.**

The aesthetic appeal of chromium metal has long been utilized in the motorcycle and automobile industry.

## CLASSIC CHROME

The familiar "chrome" that one might see on classic cars or motorcycles is in fact a thin layer of pure chromium electroplated on top of another metal, often its transition metal companion, nickel. The chrome finish was designed to be both decorative and durable, but with the advent of cheaper, equally durable composite and plastic materials, the use of chromium in the automotive industry faded after its heyday in the 1950s, '60s, and '70s.

## STAINLESS

Chromium is the metal that makes stainless steel "stainless." The addition of various percentages of chromium to steel provides the alloy with a microscopic chromium oxide layer that does not diminish the lustrous appearance, but does afford it the ability to resist corrosion.

## NOT SO SWEET

Looking back to the colorful side of chromium's history, we find two other important pigments, the chromates of lead and barium. Known as lemon yellow and chrome yellow, respectively, these compounds have a somewhat checkered past. On the plus side, the daughter of England's King George IV, Princess Charlotte of Wales (1796–1817), chose chrome yellow for the color of one of her carriages, and this helped to promote the pigment as a fashionable one of the day. Sadly, these pigments (and others) were also used as food colorings, especially in confectionery. Unfortunately, a combination of lead and chromium makes for a particularly poisonous concoction. Needless to say, the practice was soon stopped.

# Manganese

**Chemical symbol**
Mn

**Atomic number**
25

**Atomic mass**
54.938

**Boiling point**
3,742°F (2,061°C)

**Melting point**
2,275°F (1,246°C)

**Electron configuration**
2.8.13.2

**Manganese is found naturally in several minerals, often with its periodic table neighbor, iron. It is distinguished by the black color of many of its naturally occurring compounds, especially in the mineral pyrolusite, which is essentially manganese dioxide. Pyrolusite was used in glass and porcelain manufacture for adjusting color, and as a black pigment, long before anyone realized that it contained an element.**

Manganese metal was isolated in 1774, by the Swedish mineralogist Johan Gottlieb Gahn (1745–1818). Gahn's pioneering work with a blowpipe (a device used to produce more efficient reduction by directing a steady stream of air into a Bunsen burner flame) on $MnO_2$ in the form of pyrolusite and carbon in the form of charcoal produced the first pellets of the metal itself.

## SYMBOLIC AND EMBLEMATIC

Like some of its cousins among the transition metals, manganese is an essential element for humans. Most people have never come across manganese as a pure element—the metal is brittle and not particularly useful unless used in alloys. Schoolchildren sometimes encounter the substance as manganese(IV) oxide, a black, powdery compound that acts as a catalyst in the decomposition of hydrogen peroxide to produce oxygen and a compound the color of black currant juice: potassium manganate(VII), aka potassium permanganate. Generations of the same schoolchildren have mixed up the symbols for magnesium (Mg) and manganese (Mn). The manifestation of element number 25 in distinctive compounds offers much scope for confusion.

## A DIFFERENT KIND OF FILLING

There is a bizarre connection between teeth and manganese. In 1875, the HMS *Challenger* was on a scientific mission around the world. The crew recovered an enormous number of teeth belonging to the huge prehistoric shark *Megalodon* encrusted in compounds of manganese. The thickness of the manganese layers dated these teeth at millions of years old, but some others were found with much less manganese, which suggested they were only a few thousand years old. This led to wild speculation that *Megalodon* might still be lurking in the most remote depths of the oceans today.

*Megalodon* teeth found preserved in manganese compounds inspired the theory that the prehistoric shark was still living in today's oceans.

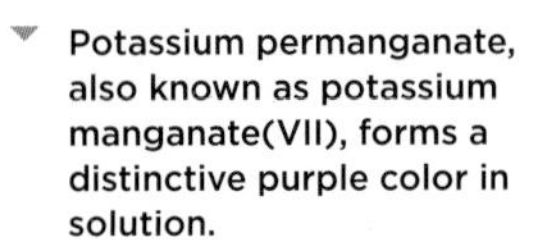

Potassium permanganate, also known as potassium manganate(VII), forms a distinctive purple color in solution.

# Iron

| | |
|---|---|
| **Chemical symbol** | Fe |
| **Atomic number** | 26 |
| **Atomic mass** | 55.845 |
| **Boiling point** | 5,182°F (2,861°C) |
| **Melting point** | 2,800°F (1,538°C) |
| **Electron configuration** | 2.8.14.2 |

**Used to create some of the earliest tools and implements, element number 26 has given its name to the period of history that started around 3,000–4,000 years ago: the Iron Age. The metal's utility has remained paramount even in modern times, especially in the ubiquitous alloy that it makes with carbon: steel.**

▲ The crucial iron-containing compound hemoglobin transports oxygen around the body, and gives red blood cells their characteristic red-pink color.

◀ Iron pyrite, a common compound of iron and sulfur, has fooled many a gold prospector over the years.

## IN THE BLOOD

Iron's role in construction is obvious, but its role at the invisible, atomic level is even more important to humans. As an essential element, iron's most crucial function within the body is to regulate oxygen transport in the blood. At the center of this crucial life-giving process is a protein called hemoglobin. This complex organic molecule has an iron(II) ion at its center, and its role is to transport oxygen from the lungs to other parts of the body. Hemoglobin is found in the red blood cells of humans, and a deficiency of iron can lead to serious medical issues. For centuries, iron tablets (usually in the form of iron(II) sulfate), have been given to patients with anemic conditions. Anemia has a number of symptoms—including fatigue—which manifest themselves when oxygen is carried inefficiently around the body.

## IRON ORES

Iron is the fourth most common element found in the earth's crust, but it is also quite abundant in the sun and stars. Iron is likely to be the element that is the most mined on the planet. As ores, it appears as hematite (iron(III) oxide) as well as the mineral pyrite, $FeS_2$. Better known as "fool's gold," its brassy yellow hue has raised the heart rate of many would-be gold hunters.

## A MAGNETIC ELEMENT

Along with cobalt and nickel, iron is one of the magnetic elements. Since the term "ferromagnetism" references iron itself, one might even consider it to be the quintessential one.

## THE RED MENACE

Iron's usefulness in construction and industry comes with a cruel, chemical twist. In most of the environments that one might expect to find iron put to use, one also finds oxygen and water. The combination of those three substances means only one thing: rust. The sight of the oh-so-familiar, reddish-brown compounds that corrode, and eventually destroy, iron objects by turning them into disintegrated, flaky masses is sometimes sad, and the process is one that humankind has been fighting for centuries.

Ferromagnetism is the physical characteristic that allows materials like iron to form permanent magnets, and to be attracted to other magnets.

# Cobalt

**Chemical symbol**
Co

**Atomic number**
27

**Atomic mass**
58.933

**Boiling point**
5,301°F (2,927°C)

**Melting point**
2,723°F (1,495°C)

**Electron configuration**
2.8.15.2

Cobalt occurs naturally only in minerals and not as the free metal. When extracted, it has a silvery appearance.

**As a shiny, silver-colored, hard metal used extensively in alloys, and as a coloring agent in the familiar blue glass, cobalt seems unremarkable among its peers in the middle of the periodic table. However, one of its isotopes, cobalt-60, has projected a scary, if unfair, reputation onto element number 27, as a potential constituent of a "dirty" nuclear bomb.**

Cobalt was discovered in a mineral found in some German mines in the 1730s by the chemist Georg Brandt (1694–1768). The name for element number 27 is derived from the mythical kobolds thought to be mischievously preventing the mineral from yielding the copper thought to be there. In fact, there was no copper present in the ore in question, but rather it was mostly made up of cobalt arsenide, $CoAs_2$. Cobalt's association with the color blue is due to its use as a pigment in painting, and in the manufacture of porcelain and glass. The chemical formula for the pigment known as cobalt blue is $CoAl_2O_4$.

For thousands of years cobalt has been used as a pigment for glass and ceramic glazes, where the element imparts a distinctive blue color.

## "DIRTY" BOMBS

There is only one naturally occurring isotope of cobalt, cobalt-59, and it is not radioactive. The nasty isotope is the synthetic cobalt-60. As a gamma ray emitter, it has a relatively long half-life, and if left unchecked has a terrifying potential to unleash massive doses of lethal radiation across large areas. As such, cobalt-60 has been discussed as a potential component of a "dirty" bomb, one where the impact of the initial blast is "enhanced" by the lingering and widespread nuclear fallout, which would leave large areas of land uninhabitable for decades.

## QUEBEC'S COBALT BEER

The historical impact of cobalt on health has not been limited to its nuclear capacity, either. In the late 1960s, a bizarre tale emerged in Canada, where excessive beer-drinking became associated with heart conditions observed in a certain group of men. Several died, and the conundrum was not solved until a common thread was found: All of the men were drinking beer from one particular brewery. When it was discovered that the brewery was adding cobalt sulfate to its beer in order to increase its head retention, cobalt poisoning was diagnosed as the culprit.

# Nickel

**Chemical symbol**
Ni

**Atomic number**
28

**Atomic mass**
58.693

**Boiling point**
5,275°F (2,913°C)

**Melting point**
2,651°F (1,455°C)

**Electron configuration**
2.8.16.2

**As one of the metals historically used to make coins, nickel has become synonymous with currency. Yet element number 28 is not a member of the group known as the "coinage metals" (that privilege is reserved for group 11, which contains copper, silver, and gold), and neither is it the main constituent of the American coin that takes its name.**

Once again, eighteenth-century German miners failed to extract copper from an ore they thought contained copper, but actually did not. Superstitiously, they blamed their failure on the devil and used the word *Kupfernickel* (literally "copper devil") to describe the bothersome, non-copper-yielding ore. When a new metal—nickel—was finally extracted from the same ore, it took its name from the original German expression of frustration.

The modern American nickel is in fact made primarily from copper and not element number 28.

## WEIRD ALLOYS

Nickel alloys exhibit some curious properties, with most of the important ones revolving around a resistance to corrosion and an ability to continue to perform at high temperatures. One such is Invar. The name is derived from the word *invariable*, a clue to its most important property. When heated, unlike other metals, Invar hardly expands *at all*. The Swiss physicist who discovered it, Charles Édouard Guillaume (1861–1938), won the 1920 Nobel Prize in Physics for "the service he has rendered to precision measurements in Physics by his discovery of anomalies in nickel steel alloys." Invar is an alloy of nickel and iron, and is one of a family of metals that also includes Elinvar. Again the name is derived from the most important property of the alloy, this time the fact that it is elastically invariable, meaning that it resists a change of shape when a force is applied. Invar and Elinvar have each found uses in instruments where precision is important: for example, in watch springs and gauges.

Another alloy of nickel, this time in an approximately 50:50 ratio with titanium and known as nitinol, has a different, incredible property. Known for being a shape-memory metal, when deformed, nitinol can spring back to its original shape. One of its most popular uses has been in the production of frames for eyewear.

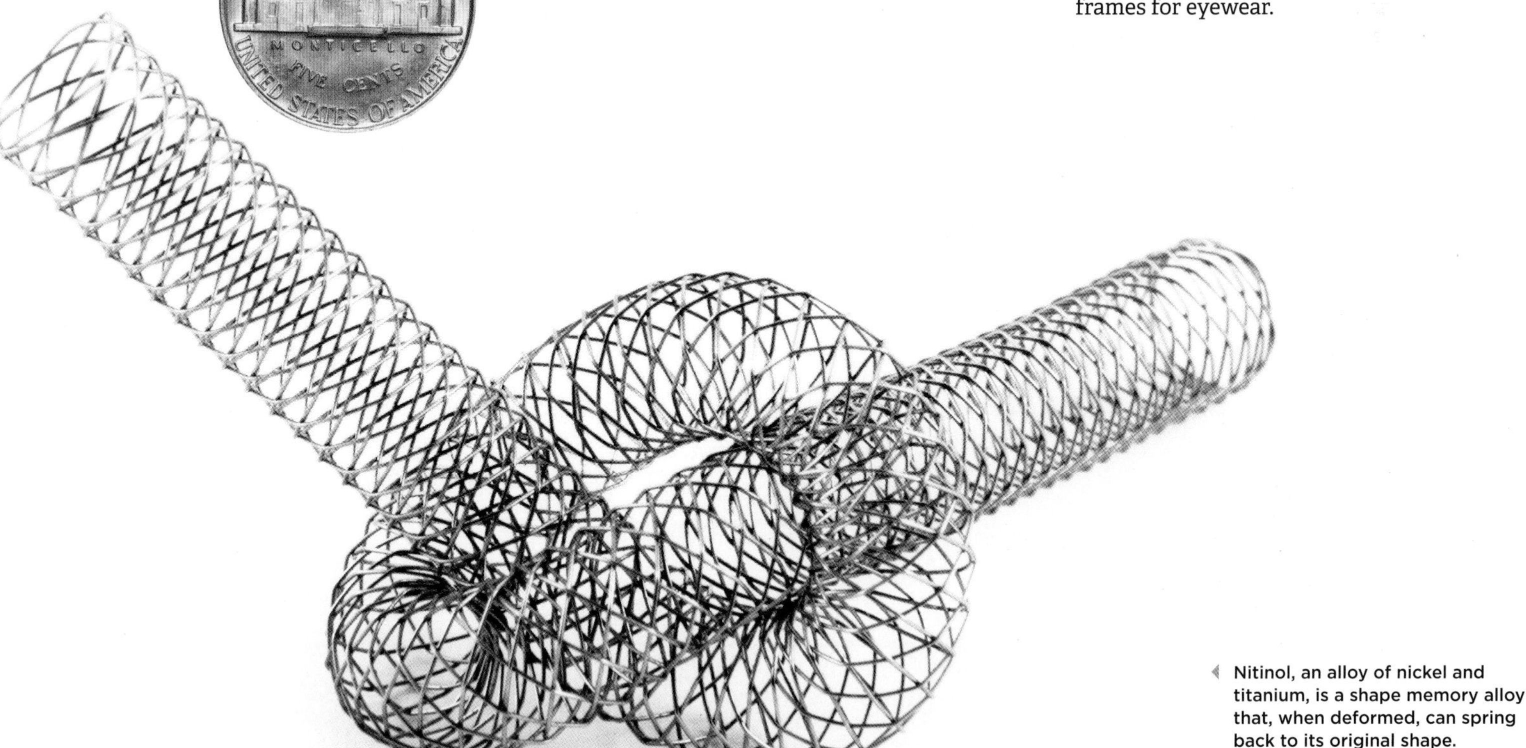

Nitinol, an alloy of nickel and titanium, is a shape memory alloy that, when deformed, can spring back to its original shape.

# Copper

| Chemical symbol | Cu |
|---|---|
| Atomic number | 29 |
| Atomic mass | 63.546 |
| Boiling point | 5,301°F (2,562°C) |
| Melting point | 1,984.32°F (1,084.62°C) |
| Electron configuration | 2.8.18.1 |

**Copper's raw color sets it apart from many of the other metallic elements, most of which are silvery, whitish, or gray. In both its pure metallic form (the familiar brownish-red color), and in the form of the vivid green patina, there is just no mistaking it.**

At the top of group 11 of the periodic table, copper is the first of the group of elements known as the coinage metals. In reality, the number of metals that are used in the manufacture of coins extends far beyond copper, gold, and silver. In terms of coins, another alloy is more important in modern times than both of those. The combination of copper and nickel (unsurprisingly known as cupronickel) has been in widespread use for coins all over the world. In the UK, coins known as "silver" for their color were actually made from the copper and nickel alloy for a number of years.

Copper's excellent conductivity has long been exploited for electrical wiring.

## CONDUCTIVITY AND RESISTIVITY

In its pure metal form, copper is an extraordinarily good conductor of electricity as well as being malleable and ductile. This combination of properties has made it a popular metal in applications such as electrical wiring. Its ability to resist attack by both water and air, along with its relatively nontoxic nature when compared to lead, means that it is now the premier metal used to make water pipes.

## BLUE-BLOODED

Copper is an essential metal for humans in homeostatic processes, and a lack of it can cause neurological disorders. In animals, copper is also essential, and in one of those essential roles it affords certain invertebrates (such as horseshoe crabs and some snails) one interesting characteristic: It makes their blood blue. The explanation lies in a biochemical molecule called hemocyanin, analogous to hemoglobin in human blood. In hemocyanin the metal at the center of the compound is not iron but copper. This important difference means that the animals have blue rather than red blood.

Some creatures, such as the horseshoe crab, have blue blood. This is caused by the hemoglobin analog hemocyanin, in which the former's iron ions are replaced by copper ions.

One of the most famous, and most visible, examples of a copper compound is found in the green patina of the Statue of Liberty.

# Zinc

| Chemical symbol | Zn |
|---|---|
| Atomic number | 30 |
| Atomic mass | 65.38 |
| Boiling point | 1,665°F (907°C) |
| Melting point | 787.15°F (419.53°C) |
| Electron configuration | 2.8.18.2 |

The German chemist Andreas Marggraf is cited as the first person to isolate a pure sample of zinc in 1746.

**As a metal, zinc is reactive, has a low melting point, and is relatively cheap, all of which add to its versatility. Whether it be the core of an American penny, or it's being used as a sacrificial anode to save another metal from corrosion, zinc gets around.**

Zinc is another element with its roots firmly in antiquity and without a specific discoverer *per se*. Known by the ancient civilizations in the form of its most famous alloy, brass, several sources cite the first isolation of the pure metal by German chemist Andreas Marggraf (1709–1782) in 1746 as the point that marks the true birth of element 30.

Zinc sits at the top of group 12, and is the final d-block element in the fourth row of the periodic table. Its classification as a transition metal is debatable. If one uses the definition of transition metals as those elements that form stable ions that have partially filled d-subshells then, like scandium, which forms an ion with an entirely empty d-subshell, zinc's 2+ ion disqualifies it, since in that ion, zinc has a completely filled d-subshell.

### GALVANIZATION

Zinc's relatively high reactivity is what makes it so useful to us. In situations where it is desirable to prevent another metal from reacting, zinc can be placed adjacent to that metal, and the zinc will react preferentially. This is exactly what happens when steel is galvanized. Not only does the layer of zinc protect the steel underneath it, but even if that protective layer is damaged and the steel exposed, it's still the zinc that will corrode before the iron in the steel.

Zinc has a typical silvery gray color. It is often found in the same deposits as lead and silver.

## REDOX REACTIONS

Zinc's ability to act as a sacrificial metal is due to a type of chemical reaction where electrons are exchanged, known as a redox reaction. Such a reaction helps to explain another popular use of zinc, as a component of batteries. All batteries work on the same principle: that one of the components will release electrons, and that the other will receive those same charged particles. As the electrons flow, electricity is generated, and the battery releases its power. In batteries, zinc, whether used in conjunction with carbon, or nickel, or even air, is the anode, where it is the species that releases the electrons that travel to the other component, known as the cathode.

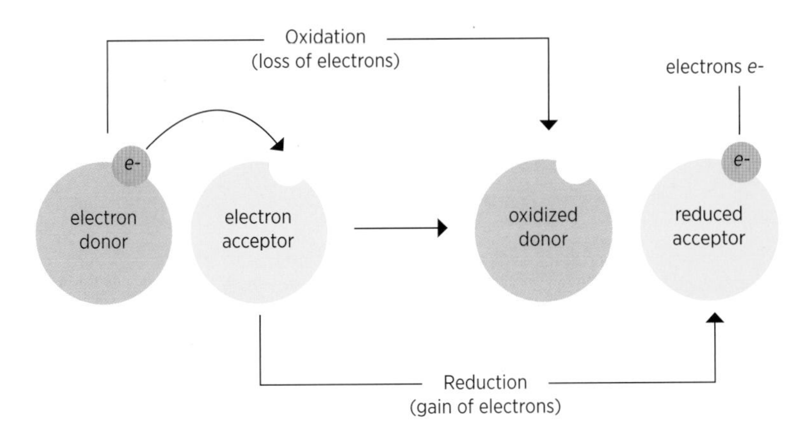


Redox reactions involve the transfer of electrons from one species to another. In the reaction of zinc metal with copper ions, the zinc transfers two electrons to the copper ions.

# Yttrium

**Chemical symbol**
Y

**Atomic number**
39

**Atomic mass**
88.906

**Boiling point**
6,053°F (3,345°C)

**Melting point**
2,771.6°F (1,552°C)

**Electron configuration**
2.8.18.9.2

Yttrium iron garnet, or YIG, has a number of special acoustic and optical applications.

**The very name of element number 39 suggests to some a bizarre, esoteric nature, but actually it can be explained quite easily. Yttrium is one of four elements named after a single, otherwise insignificant place. Why? Well, the metal is one of four elements—along with terbium (65), erbium (68), and ytterbium (70)—that were originally sourced from the mineral yttria, which was found in a mine in the small village of Ytterby, Sweden.**

## RARE OR TRANSITION?

As a transition metal (or is it?), yttrium is a dark gray solid with a melting point of over 2,771°F (1,500°C). The question as to whether yttrium should be classified as a transition metal or not is tied up in its similarity to the lanthanoids, formerly known as the rare earth elements. Ironically, most of those elements are not rare at all, but they are notoriously difficult to separate from one another, and therefore very difficult to identify.

## SUPER-COLD SUPERCONDUCTOR

Yttrium's applications are fairly specialized. For example, a compound of yttrium that includes barium, copper, and oxygen is a superconductor that works at temperatures of around –269°F (–182°C), or approximately 91 kelvin. That's considered a relatively high temperature for superconductors, with most functioning at much lower temperatures still. Such superconducting material is plunged into liquid nitrogen to achieve the necessary ultracold conditions.

## WEIRD COMPOUNDS

YIG (yttrium iron garnet) and YAG (yttrium aluminum garnet) are two pretty weird compounds that also have very specialized applications. YIG—actually a compound of yttrium, iron, and oxygen—has several interesting optical and acoustic properties, which include its ability to act as a filter for microwaves. YAG—yttrium in conjunction with aluminum and oxygen—has found use as a synthetic gemstone. It can be manufactured to resemble a diamond, and can be colored easily by the addition of other elements such as neodymium (pink) and cobalt (blue).

Neodymium-doped yttrium aluminum garnet laser crystals, or Nd:YAG, are used in medical and opthalmic lasers, and in manufacturing for engraving and cutting metals and plastics.

# Zirconium

**Chemical symbol**
Zr

**Atomic number**
40

**Atomic mass**
91.224

**Boiling point**
7,968°F (4,409°C)

**Melting point**
3,371°F (1,855°C)

**Electron configuration**
2.8.18.10.2

**Zirconium is a brutally hard metal. It is used as an industrial abrasive, but element number 40 is better known for another application: as a diamond substitute in the form of cubic zirconium.**

When German chemist Martin Heinrich Klaproth (1743–1817) was investigating a sample of jargon (a form of zircon, a diamond-like gemstone) in 1789, he extracted the oxide of a hitherto unknown element. It took a while to isolate the element itself—that didn't happen until 1824, when Jöns Jacob Berzelius (1779–1848) managed to extract it—and it even confounded the brilliant Humphry Davy and his electrolysis experiments of 1808.

A sample of zircon, essentially zirconium silicate, from which zirconium metal and zirconium dioxide can be extracted.

### CERAMIC STRENGTH

The pure element zirconium is nothing more than a typical silver-colored, shiny metal, but in many of its most important applications it looks very different. The element appears as the component of a number of ceramic materials, notably as zirconia, or zirconium dioxide, $ZrO_{2w}$. In this ceramic form, it has a number of applications related to its heat resistance and lack of chemical reactivity, such as its use in dentistry, and as a substitute for stainless steel in ceramic knives and industrial cutting tools.

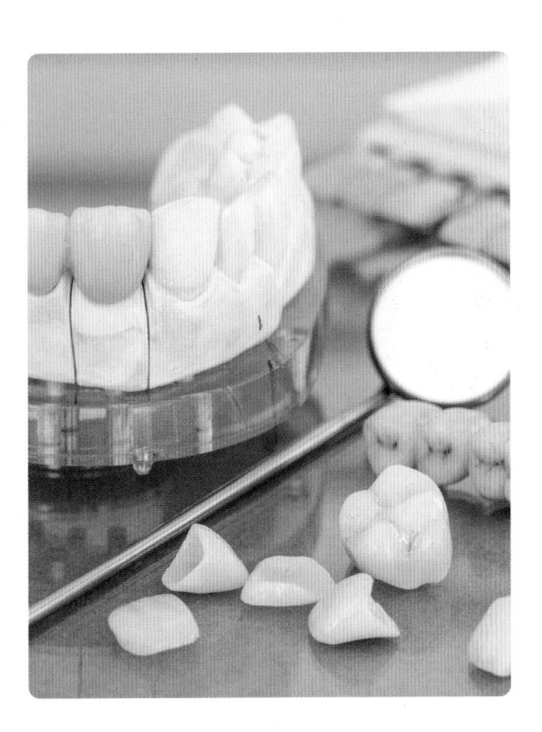

Zirconia crowns are common in modern dentistry. Known as EZR, these ceramic crowns have superior strength and are resistant to corrosion.

### COUNTERINTUITIVE

If you ask most people what happens to the physical size of objects when they are heated, most will answer that they expand. This is, of course, largely true, but there is one odd compound of zirconium that does not comply. Zirconium tungstate, $ZrW_2O_8$, has a crystalline structure that does exactly the opposite when heated: It shrinks!

### DISABLING DEADLY GAS

In 2015, scientists at Northwestern University reported an amazing property of another zirconium compound. Metal-organic frameworks (MOFs) contain metals combined with organic compounds. One particular MOF has zirconium in its structure, and the compound has been shown to disable Soman (or GD), a ferocious and deadly nerve gas. It is hoped that the technology might be put to use in disabling stockpiles of chemical weapons.

A sarin rocket being dismantled in 1990. A more modern process may use MOFs to disable chemical weapons.

# Niobium

**Chemical symbol**
Nb

**Atomic number**
41

**Atomic mass**
92.906

**Boiling point**
8,571°F (4,744°C)

**Melting point**
4,491°F (2,477°C)

**Electron configuration**
2.8.18.12.1

The anodization of niobium can produce vivid hues, a property that has been exploited in the production of these colorful Austrian coins.

**Another element with a rather complex history of discovery, and in this case, a complicated history of naming too, the controversy around element number 41 has, since 1950, settled down a bit. It was not until the middle of the twentieth century that the IUPAC formally gave it its modern name of niobium.**

## GREEK ROOTS

The element's new name is derived from Niobe, who in Greek mythology was the daughter of Tantalus. It is worth taking a quick look at the periodic table and making a note of the element directly underneath niobium: number 73, tantalum. Niobium had actually been discovered in 1801, nearly 150 years before it was named in the modern sense. Found in a mineral called columbite, the connection to tantalum is no accident, since the two elements are often found together in nature as well as lying next to one another on the periodic table.

The Large Hadron Collider (LHC) at CERN, on the border of France and Switzerland, utilizes niobium-containing magnets.

## HOT AND COLD

In a 1970 article in the American Chemistry Society's magazine *Chemistry*, niobium was described as a "space age metal." The article mentioned superconducting magnets, which have since become incredibly important in both medicine and science. Superconducting niobium-containing magnets can be found at the center of magnetic resonance imaging (MRI) used for medical diagnosis, and in particle accelerators, including the Large Hadron Collider (LHC) near Geneva, in Switzerland. In the LHC, the magnets are held at single-digit kelvin temperatures, and can produce extraordinarily strong magnetic fields, up to 2,000 times stronger than a typical household magnet.

## KALEIDOSCOPE OF COLORS

One visually stunning use of niobium has been in the manufacture of coins. The Austrian mint has produced a number of euro denomination coins with brightly colored centers surrounded by silver-colored rings. When it is anodized—a process that produces an oxide layer on the surface of the metal—varying the voltage used in the process causes different thicknesses of niobium oxides to be formed. These different thicknesses cause differing degrees of light diffraction that in turn produce some extraordinarily vivid colors.

# Molybdenum

| | |
|---|---|
| **Chemical symbol** | Mo |
| **Atomic number** | 42 |
| **Atomic mass** | 95.95 |
| **Boiling point** | 8,382°F (4,639°C) |
| **Melting point** | 4,753°F (2,623°C) |
| **Electron configuration** | 2.8.18.13.1 |

**Perhaps the element with the strangest-sounding name on the periodic table, molybdenum's moniker is derived from the Greek *molybdos*, meaning "lead." Molybdenite, the chief ore of molybdenum, is a soft, black substance similar to graphite. In fact, molybdenite's similarity to graphite is so strong that it has been used both as a lubricant and as the "lead" in pencils.**

## KEY FIGURE

### CARL WILHELM SCHEELE
1742–1786

Famously known as "hard luck Scheele," a name given to him by the author and academic Isaac Asimov, Carl Wilhelm was indeed unfortunate. Through a combination of misfortune, and not-quite-brilliant-enough chemistry, Scheele missed out on being credited with no fewer than six elements, possibly seven. The elements that slipped through his grasp were barium, chlorine, molybdenum, nitrogen, manganese, tungsten, and most notably, oxygen.

### MOLY ALLOYS

Molybdenum's chief and most important uses in the modern world revolve around the production of so-called moly steels. When alloyed with iron and carbon, molybdenum produces steels that are significantly harder and more temperature-resistant than ones that do not contain the transition element.

Molybdenum is a hard, silver-gray-colored metal that has a very high melting point. Molybdenite is the chief ore from which the element is extracted.

# Technetium

| | |
|---|---|
| **Chemical symbol** | Tc |
| **Atomic number** | 43 |
| **Atomic mass** | 98 (most stable isotope) |
| **Boiling point** | 7,709°F (4,265°C) |
| **Melting point** | 3,915°F (2,157°C) |
| **Electron configuration** | 2.8.18.14.1 |

**When Mendeleev predicted that an element (which he called eka-manganese) with an atomic weight of 100 would one day be discovered, little did he know it would prove to be a huge headache for future chemists to find. Technetium was named after the Greek *tekhnetos*, meaning "artificial," since upon its isolation in 1937 it became the first artificially manufactured element.**

It was later discovered that minuscule amounts of element 43 can be found in nature, where it is a product of the radioactive decay of naturally occurring uranium. These amounts are so small, however, and the technetium isotopes so unstable, that the amount of naturally occurring technetium is considered to be effectively zero. The element proved difficult to track down. The Italian duo of Emilio Segrè (1905–1989) and Carlo Perrier (1886–1948) finally discovered it in samples of molybdenum from the cyclotron (the circular particle accelerator) at the University of California, Berkeley.

Technetium was discovered from samples taken at the University of California, Berkeley, in 1937.

# Ruthenium

**Chemical symbol**
Ru

**Atomic number**
44

**Atomic mass**
101.07

**Boiling point**
7,502°F (4,150°C)

**Melting point**
4,233°F (2,334°C)

**Electron configuration**
2.8.18.15.1

Ruthenium is a relatively rare metal, with only a few tons produced each year. As a so-called platinum group metal, it has much in common with rhodium, palladium, platinum, iridium, and osmium.

**Ruthenium was named after the historical region of Ruthenia (part of modern-day Russia, Belarus, and Ukraine), which was home to its discoverer, Karl Karlovich Klaus (1796–1864). Some sources, however, credit Gottfried Wilhelm Osann (1796–1866) with the discovery.**

## A NEW TYPE OF "GROUP"?

The word "group" has a very specific meaning when it comes to chemistry and the periodic table, and it is a pretty simple definition to understand. A group is a vertical column in the table of elements, for example, group 17, the halogens: fluorine, chlorine, bromine, iodine, astatine, and tennessine. But ruthenium is part of a collection of elements that twists that formal designation into a new meaning. As one of the platinum group, ruthenium sits with rhodium, palladium, platinum, osmium, and iridium right at the center of the periodic table, in anything but a simple vertical column.

Ruthenium has been proposed as a replacement catalyst for iron in the Haber process, which is used to manufacture ammonia on a massive scale worldwide.

## ULTIMATE PRIZE

In 2005 ruthenium was at the center of the ultimate in chemistry fame, when Yves Chauvin (1930–2015), Robert Grubbs (b. 1942), and Richard Shrock (b. 1945) won the Nobel Prize in Chemistry "for the development of the metathesis method in organic synthesis." An important reaction in the manufacture of pharmaceuticals and polymers, the rearrangement of the organic groups surrounding the carbon-to-carbon double bonds in alkenes can be catalyzed by a ruthenium compound.

## MORE CATALYSTS

In another critically important industrial process—the manufacture of ammonia, $NH_3$, for fertilizers—a different metal, iron, is used as a catalyst. This is not a small-scale operation; in 2012 it was estimated that almost 154 million tons of ammonia were produced globally, with around 85 percent of it being used to make fertilizers. Seen in that light, catalysts are important in terms of pure economics. So what does this have to do with ruthenium? Well, it has been floated as a potential replacement for iron. The ruthenium catalysts have been found to be more efficient, but ruthenium is currently just too expensive to replace the traditional, less interesting catalyst on a large scale.

# Rhodium

| | |
|---|---|
| **Chemical symbol** | Rh |
| **Atomic number** | 45 |
| **Atomic mass** | 102.91 |
| **Boiling point** | 6,683°F (3,695°C) |
| **Melting point** | 3,567°F (1,964°C) |
| **Electron configuration** | 2.8.18.16.1 |

**Rhodium is named after the Greek *rhodon*, meaning rose-colored. William Hyde Wollaston (1766–1828) first extracted a sample of reddish-pink rhodium chloride in 1803. It was obtained from a sample of impure platinum that was dissolved in a combination of acids.**

Rhodium played a part in the development of the scanning tunneling microscope, which allows imaging at the atomic level.

## SHINY METAL

Rhodium is routinely quoted on the metal markets as one of the most expensive commodities on earth. Its scarcity is central to its high cost, but it also helps that, as a pure metal, rhodium is relatively inert and therefore super-resistant to corrosion. Because of that property it is shiny, and it is used as a reflective surface in high-performance optical mirrors.

## START ME UP

Rhodium's chemistry is largely tied up in the world of ultra-expensive catalysts. These catalytic actions are often hidden away, either in industrial plants—for example, where rhodium catalysts are used in the manufacture of nitric acid and menthol—or perhaps on the underside of your car—where rhodium is a popular metal in the somewhat more mundane catalytic converter.

# Palladium

| | |
|---|---|
| **Chemical symbol** | Pd |
| **Atomic number** | 46 |
| **Atomic mass** | 106.42 |
| **Boiling point** | 2,963°F (5,365°C) |
| **Melting point** | 1,554.9°F (2,830.82°C) |
| **Electron configuration** | 2.8.18.18 |

**It is entirely appropriate that palladium sits next to rhodium, since the pair were discovered together, by the same person, and in the same impure sample of platinum. What sets palladium apart from most other elements is the very odd way that it was announced to the world by its discoverer, and the spat it induced between him and a rival chemist.**

Experiments in cold fusion involve the electrolysis of water in a cell with a palladium cathode.

## STRANGE ANNOUNCEMENT

William Hyde Wollaston announced his discovery of palladium in a very peculiar manner. He decided to advertise the element for sale in a London shop. Wollaston wanted to profit from the sale of the metal, and by keeping it shrouded in secrecy, he thought he might have a better chance of doing just that.

Irish chemist Richard Chenevix (1774–1830) publicly chastised Wollaston and challenged the veracity of the alleged new metal. But the metal was the real deal, Wollaston won the day, and Chenevix lost all credibility in the process.

## COLD FUSION

Like rhodium, palladium's main use is in automobile catalytic converters, but it is element number 46's extraordinary interaction with hydrogen gas that is the subject of much speculation. A sample of palladium is capable of absorbing up to 900 times its own volume of hydrogen. Hydrogen storage is great, but cold fusion is where palladium's interaction with hydrogen gets interesting. This process takes place at super-high temperatures. But what if we could perform essentially the same reaction on earth, with a small electric current, so that the chemical reaction generated more energy than was used? Answer: The world's energy problems would be no more!

# Silver

| | |
|---|---|
| **Chemical symbol** | Ag |
| **Atomic number** | 47 |
| **Atomic mass** | 107.87 |
| **Boiling point** | 3,924°F (2,162°C) |
| **Melting point** | 1,763.2°F (961.78°C) |
| **Electron configuration** | 2.8.18.18.1 |

**Silver sits below copper (the first element in group 11) and above gold (the third), which positions silver firmly in second place. Despite apparently being destined for a permanent spot as a runner-up, silver does excel in some respects.**

Silver has been known to civilizations for thousands of years. In the form of coins and jewelry, it has been around since at least 3000 BCE. Like its close cousin gold, silver is a shiny metal that can be found as the free metal, but unlike gold, it rarely is.

## LET'S REFLECT FOR A MOMENT

Silver is among the most reflective of the metals, so it is often used as the reflective surface for mirrors. However, where silver takes first place hands down is in its ability to conduct electricity. The first three elements in group 11—silver, with its group mates copper and gold—are together the best conductors of electricity of all the elements in the periodic table.

## TARNISH

With several desirable properties, silver is let down in only one regard, and it is quite an important one where aesthetics are concerned: Silver tarnishes quickly, and quite badly. The buildup of the familiar black discoloration of silver sulfide on the surface of fancy cutlery and other decorative ware has certainly held silver back when compared to, say, gold. It is this tendency to corrode that also has gold beating silver in the production of electrical contacts that must stay pristine.

One area where the darkening of silver compounds has been put to constructive use is in photography, where silver ions in the +1 oxidation state, typically in silver halides such as silver bromide (AgBr), are reduced to silver metal when exposed to light, causing the darkening that creates the photographic image.

▲ Most "silver" jewelry is in fact not pure silver at all, but an alloy of silver and copper. A stamp of "925" indicates that the piece is 92.5 percent silver.

The daguerreotype photography process, developed by Louis Daguerre, used sheets of silver-plated copper. It was used for this portrait of two sisters by Antoine Claudet, taken around 1850.

## FEELING BETTER?

Silver has had an interesting role in medicine over the centuries, too, in particular as the compound silver nitrate. $AgNO_3$ has turned up in a number of medical applications, including as an antiseptic and as a wart remover. In 1999, the U.S. Food and Drug Administration effectively banned over-the-counter silver-containing products that make any claims relating to health benefits.

# Cadmium

| Chemical symbol | Cd |
|---|---|
| Atomic number | 48 |
| Atomic mass | 112.41 |
| Boiling point | 1,413°F (767°C) |
| Melting point | 609.93°F (321.07°C) |
| Electron configuration | 2.8.18.18.2 |

**A wicked element, cadmium is associated with two of its group 12 mates, zinc and mercury. Like mercury, cadmium is an insidious poison; whenever zinc is mined, you can expect some additional, unwelcome cadmium coming along for the ride.**

## OFF-COLOR

Cadmium has a rich history of use in pigments in the art world, but unfortunately its toxicity is problematic. Cadmia was a mineral used to produce zinc oxide in Germany in the nineteenth century. By heating the mineral, essentially zinc carbonate, it yields carbon dioxide and what should be a residue of white zinc oxide. Friedrich Stromeyer (1776–1835) was a bureaucrat charged with overseeing the pharmacies in Hanover at the time. He was confused over the formation of a yellowish tinge that persisted in the oxide, when the product should have been pure white. In 1817 he identified a new element, which he called cadmium after the original name for the zinc carbonate mineral.

Cadmium is often found in conjunction with zinc, so zinc mining can cause the release of toxic cadmium into the environment, as it has here, at Storey's Creek, Tasmania.

# Hafnium

| Chemical symbol | Hf |
|---|---|
| Atomic number | 72 |
| Atomic mass | 178.49 |
| Boiling point | 8,317°F (4,603°C) |
| Melting point | 4,051°F (2,233°C) |
| Electron configuration | 2.8.18.32.10.2 |

**Hafnium is named after the Latin name for Copenhagen, Hafnia, the city where it was first discovered, in 1923, by the Danish chemist Dirk Coster (1889–1950), working with his Hungarian colleague George de Hevesy (1885–1966). Hafnium was an elusive element, being one of the last of the naturally occurring elements to be discovered.**

Hafnium was one of the elements discovered after Moseley's identification of atomic numbers. The element was named after the Latin name for Copenhagen, Hafnia.

## MISSING ELEMENTS

Hafnium's discovery owes a lot to Henry Moseley's work on the atomic number. When Moseley reordered the elements via atomic number, it became apparent that a number of elements were missing, and number 72 was among those.

## LINES IN THE SPECTRUM

Hafnium atoms are almost identical to those of zirconium, element number 40. As a result, it proved difficult to isolate and find hafnium. As an element that sits on the boundary of the transition metals and the rare earth elements, the methodology best suited to discovering it was not clear. Ultimately, the element was identified via X-ray analysis, when it created lines in the spectrum that confirmed the existence of a new element. Where was the element found? Why, in a sample of zircon (zirconium silicate), of course!

# Tantalum

**Chemical symbol**
Ta

**Atomic number**
73

**Atomic mass**
180.95

**Boiling point**
9,856°F (5,458°C)

**Melting point**
5,463°F (3,017°C)

**Electron configuration**
2.8.18.32.11.2

There is a rich supply of coltan, an ore from which tantalum is extracted, in the Democratic Republic of Congo, where exploitation of the mineral has fueled conflict and led to a "coltan crisis."

**Another super-tough metal, tantalum is named after King Tantalus from Greek mythology. Its close relationship with niobium made it a tricky element to isolate. Just over a hundred years separate the first reporting of the new metal by Anders Ekeberg (1767–1813) in 1802 and its appearance as a pure element in 1903. Why such a long wait? Because separating it from the almost identical niobium proved very difficult.**

### INSIDE AND OUTSIDE THE BODY

Like titanium, tantalum's resistance to corrosion means that it can be used for the manufacture of surgical instruments that are used outside the body, and as artificial surgical implants, such as synthetic joints, that are placed inside.

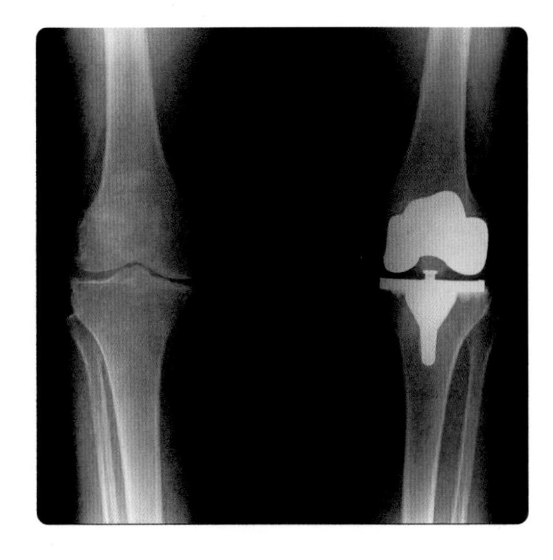

Tantalum is tough and resistant to corrosion, making it a great choice for surgical implants in the human body.

### AN ELEMENT OF CONTROVERSY

Tantalum's real power is found in its ability to conduct both electricity and heat incredibly well. This has led to its extensive use in capacitors, the tiny electronic components found in virtually every portable electronic device that anyone owns. Its durability, coupled with its superior electrical properties, means that tantalum can be made into tiny components that can store charge, but are at the same time mechanically strong enough to avoid damage.

The unprecedented popularity of small mobile electronic devices has thrust tantalum into controversy. A sign outside the gorilla enclosure at the San Diego Zoo reads thus: "Since 2004, we have recycled almost 9,000 cell phones. This reduces the demand for coltan, an ore used to make a component of cell phones, mined in gorilla habitat." "Coltan" is short for columbite–tantalite, an ore from which both niobium and tantalum are extracted. Coltan is found in various parts of the world, but one particularly rich source is located in the Democratic Republic of Congo (DRC) in central Africa. Civil war raged in the DRC from the late 1990s until 2003 (although conflict continues even today). In an attempt to fund their war efforts, multiple groups of antagonists, plus factions from neighboring countries, exploited the natural resources found in the region via an illegal trade in coltan.

# Tungsten

**Chemical symbol**
W

**Atomic number**
74

**Atomic mass**
183.84

**Boiling point**
10,031°F (5,555°C)

**Melting point**
6,192°F (3,422°C)

**Electron configuration**
2.8.18.32.12.2

**Tough, dense, and with the highest melting point among all the elements, you can find tungsten in alloys and used in places where those attributes are important. It has been used as a component of alloys on the business end of rockets, where temperatures reach several thousand degrees Fahrenheit.**

From ballpoint pens to armor-piercing ammunition, wherever super-tough materials are needed, tungsten carbide is a good choice.

## NAMES AND SYMBOLS

Perhaps the element with the most confusing chemical symbol on the periodic table, tungsten has a reason for its apparent weirdness. The symbol comes from an alternative name for element number 74, wolfram, which in turn, comes from wolframite, an ore that contains tungsten . . . or wolfram! The IUPAC's Red Book (the bible of inorganic nomenclature) has been at the center of it all, and IUPAC has some explaining to do. In 1783, the Spanish brothers Fausto (1755–1833) and Juan José Elhuyar (1754–96) had extracted tungsten from wolframite and, unsurprisingly, favored the name wolfram for the new element. Unfortunately, prior to that, Carl Wilhelm Scheele had extracted the oxide of the same element from an ore called tung sten (Swedish for "heavy stone"), so that name was also in use. For a couple of centuries both names persisted, and it was not until 1949 that the IUPAC attempted to clear up the confusion by choosing wolfram. Oddly, in 2005, the IUPAC abruptly erased all mentions of the name wolfram for element 74 from its Red Book, and tungsten replaced it.

## GOING IN HARD

Tungsten finds further fame in compounds such as tungsten carbide, whose chemical formula is WC. WC was made accidentally by French chemist Henri Moissan (1852–1907), when he heated carbon with tungsten. He had been trying to manufacture diamonds by ferociously heating iron with charcoal. However, the "diamonds" that he produced were too small to be of any commercial value. When he tried the same method with other metals, including tungsten, he inadvertently created some fragments of the new, seriously hard, compound. Years later, German and American chemists perfected the art of manufacturing tungsten carbide. It is now a staple material used for drill bits and other cutting tools. Products that have parts that need to be ultra-durable, such as the ball in a ballpoint pen, often contain it, as does armor-piercing ammunition.

Tungsten's super-high melting point has been exploited in a number of applications, including in the glowing filaments of electric lightbulbs.

# Rhenium

| | |
|---|---|
| **Chemical symbol** | Re |
| **Atomic number** | 75 |
| **Atomic mass** | 186.21 |
| **Boiling point** | 10,105°F (5,596°C) |
| **Melting point** | 5,767°F (5,596°C) |
| **Electron configuration** | 2.8.18.32.13.2 |

**Rhenium! Is it the most romantic element on the periodic table? The story of its discovery has a certain charm. It was the last of all the naturally occurring elements to be found (as long as one considers technetium to be essentially absent from the earth's crust) and shortly after, two of its joint discoverers, Walter Karl Friedrich Noddack and Ida Eva Tacke, married one another.**

### TWO AT ONE GO

At the time of the discovery of element 75, the same scientists believed that they had also discovered technetium. That discovery was eventually dismissed, but later studies suggested that they may have actually been correct, and that they should have been awarded priority for element 43 too.

### VERY DENSE

One of the rarest elements on earth, rhenium is also one of the most dense (only osmium, iridium, and platinum are more dense), and possesses one of the highest melting points (third behind only carbon and tungsten).

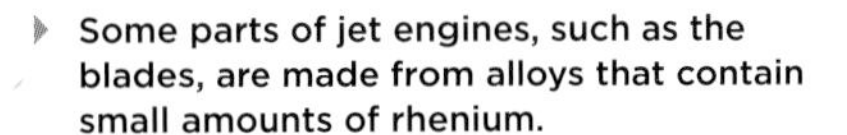

Some parts of jet engines, such as the blades, are made from alloys that contain small amounts of rhenium.

# Osmium

| | |
|---|---|
| **Chemical symbol** | Os |
| **Atomic number** | 76 |
| **Atomic mass** | 190.23 |
| **Boiling point** | 9,054°F (5,012°C) |
| **Melting point** | 5,491°F (3,033°C) |
| **Electron configuration** | 2.8.18.32.14.2 |

**Found with iridium—both physically in the earth and at the time of its discovery in 1803—osmium will always be linked to its periodic table neighbor to the right. Named after the Greek *osme*, meaning "odor" or "smell," the moniker comes from the fact that the element, along with some of its compounds, stinks!**

### GOOD SOLUTION

The English chemist Smithson Tennant (1761–1815) discovered osmium and iridium in the residue from a platinum ore dissolved in aqua regia, the same residue that led to the discovery of palladium and rhodium by William Hyde Wollaston. A combination of concentrated nitric and hydrochloric acids, it had the ability to dissolve gold metal.

### DENSE AND LIGHT?

As well as usually being recorded as the most dense of all of the elements, osmium also has a very high melting point of 5,491°F (3,033°C). With such a high melting point, element number 76 was used as a filament in the early days of incandescent lightbulbs.

Aqua regia is a concentrated solution of nitric and hydrochloric acids that gained notoriety among early chemists due to its ability to dissolve gold metal.

# Iridium

**Chemical symbol**
Ir

**Atomic number**
77

**Atomic mass**
192.22

**Boiling point**
8,002°F (4,428°C)

**Melting point**
4,435°F (2,446°C)

**Electron configuration**
2.8.18.32.15.2

The International Prototype Kilogram, used as the standard for the metric SI unit of mass, was once pure platinum but now has iridium added.

**Here's the other half of the osmium-iridium tandem. Found by Smithson Tennant in the same platinum ore residue as osmium, iridium also gets its name from one of its properties, but this time a much nicer one! Taken from the name of the Greek goddess Iris, the goddess of the rainbow, iridium was so named since its compounds can take on many colors.**

## SETTING A STANDARD

The International Prototype Kilogram, or IPK, is a cylindrical chunk of metal that defines the standard mass of one kilogram. The 1½-inch-high and 1½-inch-wide (39mm × 39mm) cylinder is kept at the International Bureau of Weight and Measures just outside Paris. The IPK, along with multiple copies, which are used for periodic verification and for practical purposes of calibration around the world, is made from a platinum and iridium alloy. The original standard kilogram was pure platinum, but iridium was added to enhance the hardness of the IPK.

## NATURAL ALLIES AND ALLOYS

Osmiridium and iridosmine are naturally occurring alloys of osmium and iridium. The names are a little misleading. Since they are naturally occurring, not only can the actual makeup of the alloys vary slightly from source to source, they usually also include other platinum group elements as part of their composition. Their superior strength and corrosion resistance is put to use in the nibs of expensive fountain pens. Because of its super-high melting point, iridium is also found in premium spark plugs; plugs made with iridium tips can outlast conventional plugs by thousands of engine miles.

## HOW HIGH CAN YOU GO?

When, in 2014, Chinese scientists managed to prepare an ion with the formula $[IrO_4]^+$, in the process they created a transition element with the oxidation state +9. A total of nine electrons had been removed from the metal, a feat never before achieved. This eclipsed the previous highest oxidation of +8, exhibited by osmium.

The use of iridium in premium-quality spark plugs has served to bring the metal to the attention of the general public.

# Platinum

**Chemical symbol**
Pt

**Atomic number**
78

**Atomic mass**
195.08

**Boiling point**
6,917°F (3,825°C)

**Melting point**
3,215.12°F (1,768.4°C)

**Electron configuration**
2.8.18.32.17.1

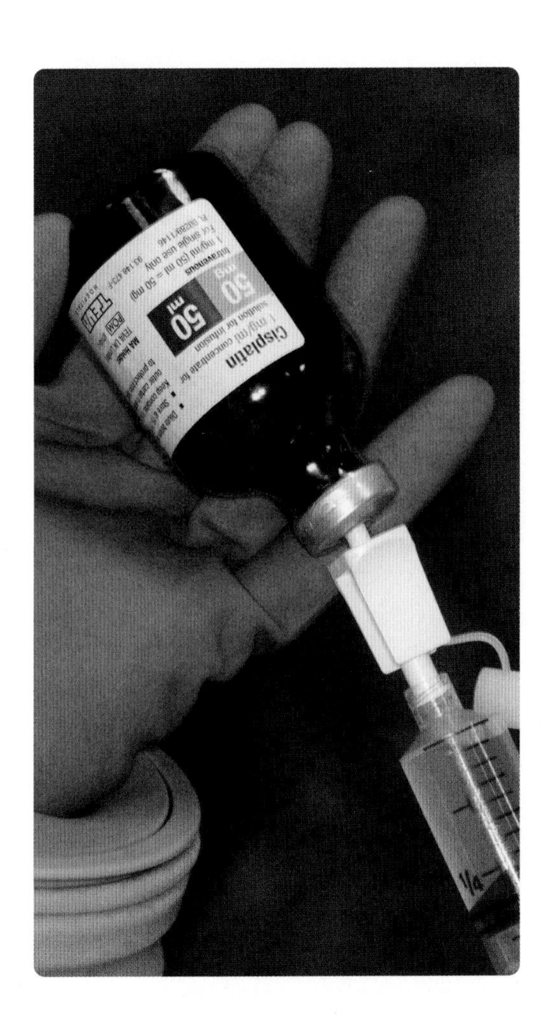


Cisplatin is a platinum-based chemotherapy drug used in the treatment of many cancers, as it can help to arrest the growth of tumors.

**As far as the precious metals go, platinum sits at the top, beating gold and silver hands down in most hierarchies. Consider, for example, the platinum disk awarded to U.S. recording artists who sell one million copies of an album, compared to the mere 500,000 required for gold; silver doesn't even register on the scale. Where silver does enter the platinum equation is via the name: *Platinum* is derived from the Spanish *platina*, meaning "little silver."**

## PRICEY

Platinum's usefulness and position in the pecking order relates to its super-resistance to reactions with just about anything. As a silvery-white metal, it has found favor as a popular choice for jewelry. It is not particularly widespread, being only the seventy-fifth most abundant element in the earth's crust. It is also reasonably difficult to extract and work with, and coupled with a high demand, these properties make platinum one of the most expensive metals in the world. As such, one might think that historically, platinum would have been a popular choice in the manufacture of coins. One notable example is the production of Russian rubles in the two decades prior to 1846. Around 1819, significant amounts of the metal were discovered in the Ural goldfields around the city of Ekaterinburg. This led to the minting of over one million coins. These coins have become collectors' items, fetching, in some cases, tens of thousands of dollars.

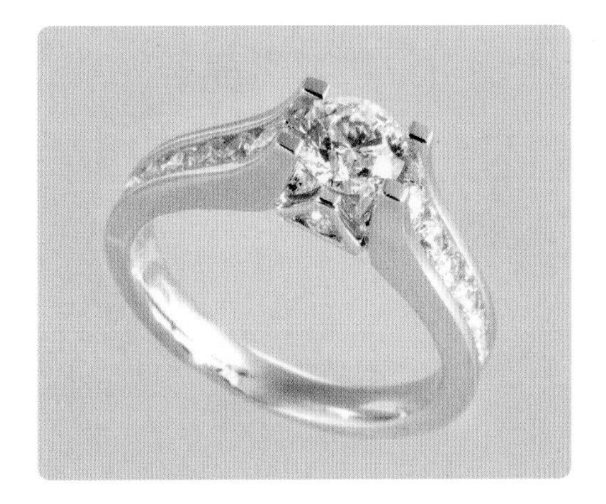

## PLATINOSIS AND ANTI-CANCER DRUGS

Although it sounds a little unlikely, platinosis—an allergic reaction to platinum salts—is indeed a real medical condition. It usually manifests itself via irritation of the upper respiratory system, and it is thought of as an occupational condition brought on by chronic exposure to platinum-based compounds. Since the late 1970s, several platinum-based drugs have been used as anti-cancer medicines; Cisplatin inhibits cell division, and has been used as an incredibly successful anti-cancer drug.

## MORE CATALYSTS, MORE ALLOYS, MORE PRIZES

When used as an alloy and a catalyst, platinum exhibits many typical transition metal traits. In fact, the Nobel Prize in Chemistry in 2007 was awarded to Gerhard Ertl (b. 1936) for his work on "chemical processes on solid surfaces." That work included the action of platinum in the catalytic converter in turning the unwanted carbon monoxide in the exhaust gases of automobiles into the less poisonous, but still not entirely harmless, carbon dioxide.

This platinum medal of 1826 shows Tsar Nicolas I of Russia. The use of platinum in Russia reached a peak in the early part of the nineteenth century.

Like silver, most platinum jewelry is less than 100 percent pure. The metal's inertness and exclusivity make it a popular choice for wedding and engagement rings.

# Gold

| Chemical symbol | Au |
|---|---|
| Atomic number | 79 |
| Atomic mass | 196.97 |
| Boiling point | 5,173°F (2,856°C) |
| Melting point | 1,947.52°F (1,064.18°C) |
| Electron configuration | 2.8.18.32.18.1 |

**Gold is one of the most iconic elements, a symbol of opulence and wealth. Its property of being relatively inert—and therefore able to resist corrosion—contributes to its ornamental appeal, and also means gold has an important role in electronic applications where excellent, long-lasting electrical contact is crucial.**

Gold has an important role in the history of chemistry, if for no other reason than that it partly drove the search for the philosopher's stone. Alchemists, as the pioneering chemists, had gold at the center of their thoughts as they sought methods to convert base metals such as lead into the more valuable and prestigious ones. Many of those efforts laid the groundwork for what we know now as modern, quantitative, and highly analytical chemistry, and gold was the enticing element.

▼ The search for the philosopher's stone and the quest for gold drove the alchemists, and, it can be argued, gave birth to much of modern chemistry in the process.

Gold plays a role as a conductor in electronics, and there are significant amounts of the valuable metal in the tons of electrical goods that are discarded each year.

## A COVETED ELEMENT

Producing coins from gold makes sense, since there is a certain intrinsic value to element number 79; however, even as a relatively rare (it is the seventy-third most abundant in the earth's crust) and relatively expensive metal, there are several others that beat it in both categories. So what is it about gold that makes it so irresistible? It is probably due to its unique color and its tremendous ability to resist corrosion over time, and hence maintain its distinctive shimmer. In fact, as it attracted the gaze of the earliest civilizations, that shine is what led gold to become one of the first known elements.

## MINING FOR GOLD IN ELECTRONICS

Gold's resistance to corrosion is not only an aesthetic attribute. In situations where it is absolutely vital for the conductivity of an electronic device to be maintained over time, gold is a popular choice for the contacts. In fact, the extensive use of gold in electronic devices has driven some people to attempt to recover it from discarded phones, cameras, and laptops. The process is tedious and hazardous, but that hasn't put people off.

## NANOMEDICINE

In 2012, scientists reported some success in treating prostate cancer with the use of gold-198 nanoparticles. The beta particles emitted from gold-198 have a range of approximately ⅜ inch (1cm), so gold atoms inserted inside a tumor could attack the offending cells without going on to damage other nearby healthy cells.

# Mercury

| | |
|---|---|
| **Chemical symbol** | Hg |
| **Atomic number** | 80 |
| **Atomic mass** | 200.59 |
| **Boiling point** | 6,74.11°F (356.73°C) |
| **Melting point** | -37.89°F (-38.8°C) |
| **Electron configuration** | 2.8.18.32.18.2 |

**The adjective *mercurial* has its origins in Roman mythology and the messenger of the gods, Mercury. However, anyone who has seen his namesake element escape the confines of an old-fashioned thermometer and skate across the floor will recognize how appropriate it is that the metal, too, has come to be associated with all things erratic and changeable.**

### ANCIENT ORIGINS

Mercury's main ore, cinnabar, was known to ancient civilizations across the world long ago. Coveted for its brilliant scarlet color, the rock—which is formed of the compound mercury(II) sulfide—was used to produce the red pigment known as vermilion.

A known toxin, mercury is associated with a wide range of health hazards. Whether it is connected with the poisoning of the developing fetus, or damage to the nervous, digestive, and immune systems, it has caused sufficient concern to create a global initiative intended to reduce its buildup in the environment: the Minamata Convention on Mercury. One of the things addressed by the convention is the bioaccumulation of methylmercury in fish. As the organometallic ion $[CH_3Hg]^+$, naturally occurring mercury in water can be concentrated in organisms and can ascend the food chain.

Cinnabar, essentially mercury(II) sulfide, was used for centuries to produce pigments and for making jewelry.

The legendary liquid metal also known as quicksilver has proved to be a fascinating element for centuries.

▲ Milliners used mercury(II) nitrate to smooth out animal pelts for fur hats. The health hazards for the hatmakers were considerable.

## MILLINERY AND MADNESS

The commonly used phrase "as mad as a hatter" has its origins in the use of the nefarious heavy metal. In the nineteenth century, milliners used mercury to smooth out animal pelts in a process called "carroting," where mercury(II) nitrate helped to mat the fine animal hairs together. Many hatmakers developed "hatter's shakes"—a condition that caused them to tremble—and other abnormal neurological conditions.

Nowadays, mercury is used largely in industrial settings as a catalyst, for example, in the production of vinyl chloride monomer (VCM), a precursor to the more familiar polyvinyl chloride (PVC).

# POST-TRANSITION METALS

Also known as the "poor" metals, these seven elements are certainly metallic in nature; they just aren't very good at being metals! Characterized by their relatively low melting and boiling points, poor strength, and brittleness, their utility is often found in their use in alloys. When mixed with other, "better" metals, properties that might otherwise be considered shortcomings can be utilized to manipulate the density, melting point, and strength of composite materials.

On the following pages:

| | | | |
|---|---|---|---|
| **Al** | Aluminum | **Tl** | Thallium |
| **Ga** | Gallium | **Pb** | Lead |
| **In** | Indium | **Bi** | Bismuth |
| **Sn** | Tin | | |

| 1 | 2 | 3 | 4 | 5 | 6 | 7 | 8 | 9 | 10 | 11 | 12 | 13 | 14 | 15 | 16 | 17 | 18 |
|---|---|---|---|---|---|---|---|---|---|---|---|---|---|---|---|---|---|
| 1 H | | | | | | | | | | | | | | | | | 2 He |
| 3 Li | 4 Be | | | | | | | | | | | 5 B | 6 C | 7 N | 8 O | 9 F | 10 Ne |
| 11 Na | 12 Mg | | | | | | | | | | | **13 Al** | 14 Si | 15 P | 16 S | 17 Cl | 18 Ar |
| 19 K | 20 Ca | 21 Sc | 22 Ti | 23 V | 24 Cr | 25 Mn | 26 Fe | 27 Co | 28 Ni | 29 Cu | 30 Zn | **31 Ga** | 32 Ge | 33 As | 34 Se | 35 Br | 36 Kr |
| 37 Rb | 38 Sr | 39 Y | 40 Zr | 41 Nb | 42 Mo | 43 Tc | 44 Ru | 45 Rh | 46 Pd | 47 Ag | 48 Cd | **49 In** | **50 Sn** | 51 Sb | 52 Te | 53 I | 54 Xe |
| 55 Cs | 56 Ba | 57-71 | 72 Hf | 73 Ta | 74 W | 75 Re | 76 Os | 77 Ir | 78 Pt | 79 Au | 80 Hg | **81 Tl** | **82 Pb** | **83 Bi** | 84 Po | 85 At | 86 Rn |
| 87 Fr | 88 Ra | 89-103 | 104 Rf | 105 Db | 106 Sg | 107 Bh | 108 Hs | 109 Mt | 110 Ds | 111 Rg | 112 Cn | **113 Nh** | **114 Fl** | **115 Mc** | **116 Lv** | 117 Ts | 118 Og |
| | | | 57 La | 58 Ce | 59 Pr | 60 Nd | 61 Pm | 62 Sm | 63 Eu | 64 Gd | 65 Tb | 66 Dy | 67 Ho | 68 Er | 69 Tm | 70 Yb | 71 Lu |
| | | | 89 Ac | 90 Th | 91 Pa | 92 U | 93 Np | 94 Pu | 95 Am | 96 Cm | 97 Bk | 98 Cf | 99 Es | 100 Fm | 101 Md | 102 No | 103 Lr |

The aluminum alloy magnalium can be used in handheld sparklers to produce bright white sparks.

## IN THE MIX

Classic examples of transition metal alloys include solder (tin and lead), bronze (copper and tin), and a whole host of aluminum alloys with evocative names that sometimes give away their composition, like magnalium (magnesium and aluminum), and are sometimes less obvious, like duralumin, which contains copper, magnesium, and manganese, along with aluminum.

## PERIODIC TRENDS

Found on the periodic table in groups 13, 14, and 15, the metals are known as "post-transition": They appear after the transition metals in groups 4 through 12. I have chosen aluminum, gallium, indium, tin, thallium, lead, and bismuth to be part of this collection, although different sources sometimes include other elements, including those from groups 11 and 12 to the left, and 16 and 17 to the right. The recent addition to the periodic table of elements 112 through 117 also provokes interesting conversations about their ultimate classification.

The position of the post-transition metals on the periodic table is worth careful consideration. With the archetypal metals to their left, and the metalloids (whose metallic properties are increasingly diluted) to their right, the post-transition metals serve as a useful reminder of how periodic trends in the table work. With "no-doubt" metals on the far left of the table, and "no-doubt" nonmetals on the far right, as one traverses the table from left to right, one can expect a gradual metamorphosis of metallic to nonmetallic behavior. The gap between the two extremes is bridged first by the poor metals, and then by the metalloids—which, perhaps by the same token, might collectively be renamed the "poor nonmetals."

Lead, together with tin, forms part of the important alloy solder, which is used for joining electrical components.

# Aluminum

| Chemical symbol | Al |
|---|---|
| Atomic number | 13 |
| Atomic mass | 26.982 |
| Boiling point | 4,566°F (2,519°C) |
| Melting point | 1,220.58°F (660.32°C) |
| Electron configuration | 2.8.3 |

**Aluminum may now be ubiquitous, and relatively cheap, but this wasn't always the case. In the first few decades after its discovery, aluminum was regarded as one of the most precious of all metals. It was treated as such, and revered, until a young Frenchman and a young American perfected a way to produce it for a sensible price.**

### FROM RARE TO OMNIPRESENT

Despite somewhat successful efforts by the French chemist Henri Sainte-Claire Deville (1818–81) to reduce the cost of manufacturing element number 13, it was still financially out of reach. The story goes that another Frenchman, Paul Héroult (1863–1914), and an American, Charles Hall (1863–1914), each independently read a book Deville had written on aluminum production, and both men were inspired to seek a more commercially viable way to obtain the light but strong metal. Héroult and Hall worked independently on perfecting an electrolytic process, and each came to almost the same conclusion, via almost the same process, at almost the same time. The so-called Hall-Héroult process revolutionized the production of aluminum, and the metal became the popular material it is today.

Aluminum is the most abundant metal in the earth's crust, and the third most abundant element overall. Like the other poor metals, on its own aluminum is quite weak and relatively soft. However, as an alloy, its low density becomes a huge asset. The metal is extraordinarily workable, meaning that it can be drawn into wires (ductility), shaped (malleability), smelted and cast, as well as rolled into sheets and even produced in powder form. This versatility means that aluminum has found uses in a huge number of areas, ranging from construction, to household cookware, to the manufacture of food cans. It also plays a vital role in the aerospace and aircraft industry, where its high strength-to-weight ratio makes it a hugely important component of a large number of alloys.

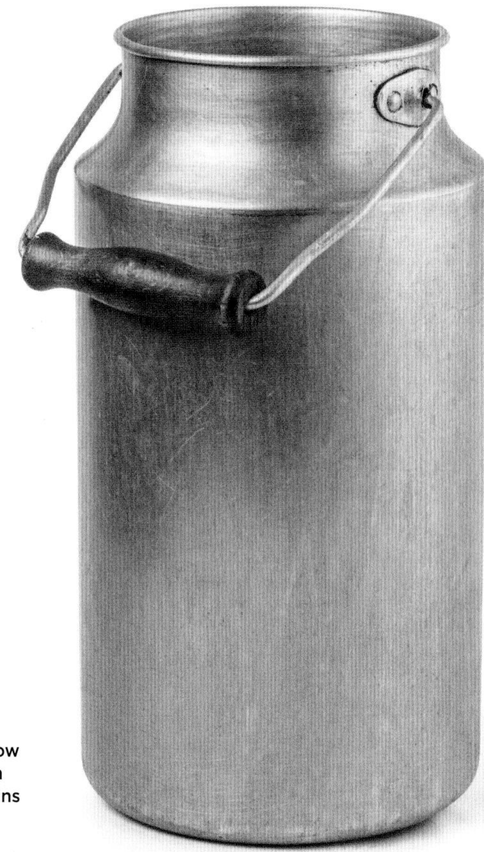

Aluminum has many applications thanks to its strength coupled with its low density. For example, when filled, this milk churn remains light enough to carry.

A very workable metal, aluminum can be machined into extremely thin sheets before being fabricated.

# Gallium

| | |
|---|---|
| **Chemical symbol** | Ga |
| **Atomic number** | 31 |
| **Atomic mass** | 69.723 |
| **Boiling point** | 3,999°F (2,204°C) |
| **Melting point** | 85.57°F (29.76°C) |
| **Electron configuration** | 2.8.18.3 |

**Gallium's claim to fame is that it is *almost* a liquid at room temperature. The fact that it melts at around 86°F (30°C) means that unlike mercury, whose unique appearance as a liquid metal at room temperature is obvious, gallium usually has the appearance of a solid, shiny metal.**

## PREDICTING PERIODICITY

Mendeleev had predicted the existence of gallium long before it was actually found or named. Mendeleev came to the conclusion that his nascent arrangement of elements must allow for the ultimate insertion of some as yet undiscovered elements. These would account for some apparent gaps in the periodic pattern he had derived. Mendeleev predicted that an element he called eka-aluminum would have properties that included an atomic mass of 68, a density of 6.0 g/cm$^3$, and a low melting point.

When, in 1875, Frenchman Paul-Émile (François) Lecoq de Boisbaudran (1838–1912) discovered a new element, he found its properties were astonishingly similar to Mendeleev's eka-aluminum. Such was the accuracy of Mendeleev's predictions that the discovery of gallium helped to solidify the whole idea of the periodic table and its patterns beyond any reasonable doubt.

Unusually for a metal, elemental gallium expands when it solidifies, meaning that, like water, its solid form will float in its own liquid.

Gallium's famously low melting point is often highlighted by the fact that the warmth from one's hand is sufficient to melt the metal.

# Indium

**Chemical symbol**
In

**Atomic number**
49

**Atomic mass**
114.82

**Boiling point**
3,762°F (2,072°C)

**Melting point**
313.88°F (156.6°C)

**Electron configuration**
2.8.18.18.3

**Indium was discovered in 1863 by the German pair Ferdinand Reich (1799–1882) and Hieronymus Theodor Richter (1824–98). While working at the Freiberg School of Mines, Reich noticed something odd about a yellow solid he had produced when experimenting with a mineral form of zinc. When the new material was interrogated, it revealed a hitherto unknown bluish-purple line in its spectrum. They christened the new metal after the Latin *indicium*, meaning "indigo," in recognition of the colored line they had observed.**

## LEAD-LIKE

Indium resembles many of the elements around it on the periodic table, not only in group 13 (Ga and Ti) as one would expect, but also diagonally down to its right, where we find lead. Like lead, indium is a very soft metal, with a value of 1.2 on the Mohs scale of hardness compared to lead's 1.5. It has a high density (7.31 g/cm$^3$ versus 11.34 g/cm$^3$ for lead) and it melts at a relatively low temperature of 314°F (157°C) compared to lead's 621°F (327°C).

## METAL MIMIC

Indium is commonly found in a number of naturally occurring zinc compounds. This curiosity is linked to the relative sizes of the $Zn^{2+}$ ion and the $In^{3+}$ ion, indium's most common oxidation state (the charge on the ion). The zinc ion has a radius of 88 picometers (one picometer is equal to $1 \times 10^{-12}$ meters), and that of the indium ion is 94 picometers. As with thallium and potassium, this kind of similarity can lead to one ion mimicking another, and explains why indium is so often found associated with ores that contain $Zn^{2+}$.

## THE LIQUID MIRROR

Indium's alloys are similar to others of the post-transition metals inasmuch as they often have very low melting points. Mixed with its fellow group 13 family member and fellow low-melting-point metal gallium, indium (along with a little tin) can produce an alloy that is liquid at room temperature. The resultant alloy, galinstan, is a silver-colored liquid that can be applied to glass to create a mirrored surface. Another of indium's alloys, indium tin oxide, is used to produce highly conductive, transparent coatings for touch screens, flat-screen displays, and solar cells.

◀ Solar cells are often coated with thin films of indium alloys, such as copper indium gallium selenide, which is is an excellent absorber of sunlight.

# Tin

**Chemical symbol**
Sn

**Atomic number**
50

**Atomic mass**
118.71

**Boiling point**
4,716°F (2,602°C)

**Melting point**
449.47°F (231.93°C)

**Electron configuration**
2.8.18.18.4

**Tin is one of the ancient elements. Arguably, its greatest contribution to civilization is as a component of one of the most historically important alloys ever known: bronze. As early civilizations moved from stone to metals to make implements and weapons, tin was mixed with copper to produce the alloy. Bronze was so critical to the development of human culture that a whole period of civilization was named for it.**

These Bronze Age implements represent an important milestone in human deveopment: the move from using tools made of stone to those made of metal.

## THE "TIN" CAN

The Latin name for tin was *stannum*, hence the somewhat confusing chemical symbol for tin, Sn. Tin is relatively soft with a low melting point. It is a silvery-white metal resistant to corrosion in a similar way to aluminum: They both form a protective oxide layer on the surface of the metal. This property was utilized in one of its largest applications, the "tin can." Even when first invented in the early part of the nineteenth century, the tin can was not made exclusively from tin; in fact, it contained very little tin at all. Most cans were made from steel, with a thin coating of tin applied to prevent rusting. Tin's resistance to corrosion meant that even the acidic nature of many foods would not damage the can, and goods could be preserved for long periods. Modern "tin" cans are often made from aluminum, which offers similar resistance to corrosion but is lighter and cheaper.

Tin's use in canning is also partly due to the fact that in many compounds tin is nontoxic.

"Tin" cans have evolved in terms of their composition and construction over the years, but they were never made purely from tin.

# Thallium

| | |
|---|---|
| **Chemical symbol** | Tl |
| **Atomic number** | 81 |
| **Atomic mass** | 204.38 |
| **Boiling point** | 2,683°F (1,473°C) |
| **Melting point** | 579°F (304°C) |
| **Electron configuration** | 2.8.18.32.18.3 |

**Thallium's discovery was subject to much debate. Sitting horizontally on the periodic table between a couple of metals, mercury and lead, thallium is a soft, silver-white metal that typically forms two ions, $Tl^{+}$ and $Tl^{3+}$.**

## CONTROVERSY FROM BIRTH

In 1861, Englishman William Crookes (1832–1919) observed the distinctive green line in the spectrum of thallium, which led to his claiming priority for its discovery. However, things were not quite so simple, largely due to the almost simultaneous work done by Frenchman Claude-Auguste Lamy (1820–78). Lamy also observed the spectral lines that Crookes had seen, and he too laid claim to the discovery. Lamy's claim was cemented by the fact that he also managed to isolate the pure element in 1862. It was not until Crookes became a Fellow of the Royal Society in 1863, and pulled rank, that the controversy was settled: Crookes was given the recognition.

Elemental thallium held under an argon atmosphere. Thallium is a soft, malleable, silver-colored metal that discolors rapidly in air.

# Lead

| | |
|---|---|
| **Chemical symbol** | Pb |
| **Atomic number** | 82 |
| **Atomic mass** | 207.2 |
| **Boiling point** | 3,180°F (1,749°C) |
| **Melting point** | 621.43°F (327.46°C) |
| **Electron configuration** | 2.8.18.32.18.4 |

**Known for its softness, malleability, and high density, lead is one of the more familiar metallic chemical elements. But while it may be familiar, it is not well thought of in the twenty-first century. The need to remove lead from paints and gasoline, and to prevent it from tainting the water supply, has harmed the element's public image.**

## SO CIVILIZED

The use of lead dates back several thousand years. Lead water pipes were used extensively by the ancient Romans because of the element's durability and malleability, and plumbing was still being made of lead well into the twentieth century. (As lead's toxic nature was understood, lead pipes were phased out in new construction.) Its use as cookware and at the heart of the water supply is thought to have contributed to the decline of the Roman Empire due to its effect on fertility. Even so, the Roman influence on the history of lead remains indelible, with the symbol for element number 82 derived from the Latin for lead, *plumbum*.

Lead was used to make water pipes for plumbing for thousands of years. This lead pipe was used to deliver water to Roman baths in Bath, England.

# Bismuth

**Chemical symbol**
Bi

**Atomic number**
83

**Atomic mass**
208.98

**Boiling point**
2,847°F (1,564°C)

**Melting point**
520.52°F (271.4°C)

**Electron configuration**
2.8.18.32.18.5

**As a neighbor of lead, bismuth has spent most of its history in the shadow of the element to its left, not least because it was confused with lead for much of its early existence. Bismuth resembles lead in many of its properties, and when the science of chemical analysis was in its infancy, the confusion with lead was inevitable. Known now as a distinct element, bismuth is still a little obscure, finding very few common applications.**

It wasn't until 1753, when French chemist Claude François Geoffroy (1729–53) published his observations in *Mémoires de l'académie française*, that bismuth became recognized as a distinct element in its own right.

The relatively low melting point of Wood's metal, an alloy of bismuth, allows plugs in sprinkler systems to melt, releasing water from the extinguishers.

## WOOD'S METAL

Bismuth's comparisons to other elements do not end with lead. Much of its chemistry is similar to other elements in the group 15 family, notably antimony and arsenic. Bismuth is also used in alloys—the quintessential application for so many poor metals. The bismuth alloy of primary importance is known as Wood's metal, an alloy developed by Barnabas Wood (1819–75), an American dentist. His alloy, made from about 50 percent bismuth with lead, tin, and cadmium, had a melting point of only 158°F (70°C). This tendency to melt at a low temperature has two applications that have helped to make the world a safer place. As a component of fire sprinklers, Wood's metal is used as a plug that seals a water pipe. As a fire burns, the temperature around the plug increases, eventually melting the metal plug, and releasing water. In a similar application, gas canisters can be fitted with valves made of Wood's metal, so that, in a fire, the canister will open, releasing the gas inside and thus reducing the risk of a cylinder exploding.

## HEARTBURN AND NUCLEAR FUSION

A more familiar application of bismuth is that of one of its compounds, bismuth subsalicylate. Under the commercial name Pepto-Bismol, the bright pink liquid is sold as an over-the-counter medicine used to treat mild stomach ailments.

Going from self-medication to some of the most extreme chemistry ever conducted—nuclear fusion—makes for an interesting juxtaposition, but bismuth has played a role in both. As a target nucleus, bismuth-209 was bombarded with iron-58 atoms and the result was a new element. In 1982, a single, short-lived atom of element number 109, meitnerium-266, was produced in this way by a team led by Peter Armbruster (b. 1931) and Gottfried Münzenberg (b. 1940) in Germany.

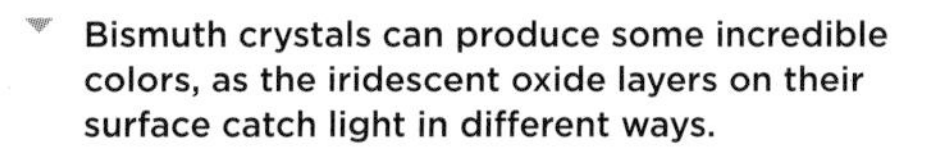

Bismuth crystals can produce some incredible colors, as the iridescent oxide layers on their surface catch light in different ways.

The familiar pink color of the over-the-counter upset stomach remedy Pepto-Bismol. The name gives away the presence of element number 83.

# METALLOIDS

The metalloids are an eclectic bunch. Not only does this ragtag collection of elements exhibit chemical and physical behaviors that cut across the boundaries between metals and nonmetals, they are also found scattered between groups 13 and 17 of the periodic table. Consequently, they have varying numbers of electrons in their valence shells, and since valence electrons drive chemical behavior, these elements exhibit a wide variety of chemical properties.

On the following pages:

| | | | |
|---|---|---|---|
| **B** | Boron | **Sb** | Antimony |
| **Si** | Silicon | **Te** | Tellurium |
| **Ge** | Germanium | **Po** | Polonium |
| **As** | Arsenic | | |

| 1 | 2 | 3 | 4 | 5 | 6 | 7 | 8 | 9 | 10 | 11 | 12 | 13 | 14 | 15 | 16 | 17 | 18 |
|---|---|---|---|---|---|---|---|---|---|---|---|---|---|---|---|---|---|
| 1 H | | | | | | | | | | | | | | | | | 2 He |
| 3 Li | 4 Be | | | | | | | | | | | **5 B** | 6 C | 7 N | 8 O | 9 F | 10 Ne |
| 11 Na | 12 Mg | | | | | | | | | | | 13 Al | **14 Si** | 15 P | 16 S | 17 Cl | 18 Ar |
| 19 K | 20 Ca | 21 Sc | 22 Ti | 23 V | 24 Cr | 25 Mn | 26 Fe | 27 Co | 28 Ni | 29 Cu | 30 Zn | 31 Ga | **32 Ge** | **33 As** | 34 Se | 35 Br | 36 Kr |
| 37 Rb | 38 Sr | 39 Y | 40 Zr | 41 Nb | 42 Mo | 43 Tc | 44 Ru | 45 Rh | 46 Pd | 47 Ag | 48 Cd | 49 In | 50 Sn | **51 Sb** | **52 Te** | 53 I | 54 Xe |
| 55 Cs | 56 Ba | 57-71 | 72 Hf | 73 Ta | 74 W | 75 Re | 76 Os | 77 Ir | 78 Pt | 79 Au | 80 Hg | 81 Tl | 82 Pb | 83 Bi | **84 Po** | 85 At | 86 Rn |
| 87 Fr | 88 Ra | 89-103 | 104 Rf | 105 Db | 106 Sg | 107 Bh | 108 Hs | 109 Mt | 110 Ds | 111 Rg | 112 Cn | 113 Nh | 114 Fl | 115 Mc | 116 Lv | 117 Ts | 118 Og |
| | | | 57 La | 58 Ce | 59 Pr | 60 Nd | 61 Pm | 62 Sm | 63 Eu | 64 Gd | 65 Tb | 66 Dy | 67 Ho | 68 Er | 69 Tm | 70 Yb | 71 Lu |
| | | | 89 Ac | 90 Th | 91 Pa | 92 U | 93 Np | 94 Pu | 95 Am | 96 Cm | 97 Bk | 98 Cf | 99 Es | 100 Fm | 101 Md | 102 No | 103 Lr |

## WHAT ARE THEY, ANYWAY?

Some of the elements that are usually included in this assemblage are made up of relatively small, light atoms, whereas others have very heavy atoms. The size and mass variation adds to the complexity of their different behaviors, and settling on any universal description for the metalloids as a whole is virtually impossible. The term "metalloid" itself has had a number of different meanings. The modern meaning (that of exhibiting the properties of both metals and nonmetals) is still somewhat ill defined.

It is probably not surprising, then, given the difficult history associated with even the word itself, that chemists cannot completely agree on which elements should be considered metalloids. Elements that one might normally consider to be firmly in the realm of metals—such as zinc, tin, chromium, and lead—and those that are commonly considered nonmetals without much dispute—like nitrogen, sulfur, and phosphorous—have, at one time or another, been classified as metalloids. In modern chemistry the elements that are most commonly thought of as metalloids are boron, silicon, germanium, arsenic, antimony, and tellurium, with polonium and astatine sometimes joining those six. In this book, we will consider the most common half-dozen plus polonium to be "our" metalloids.

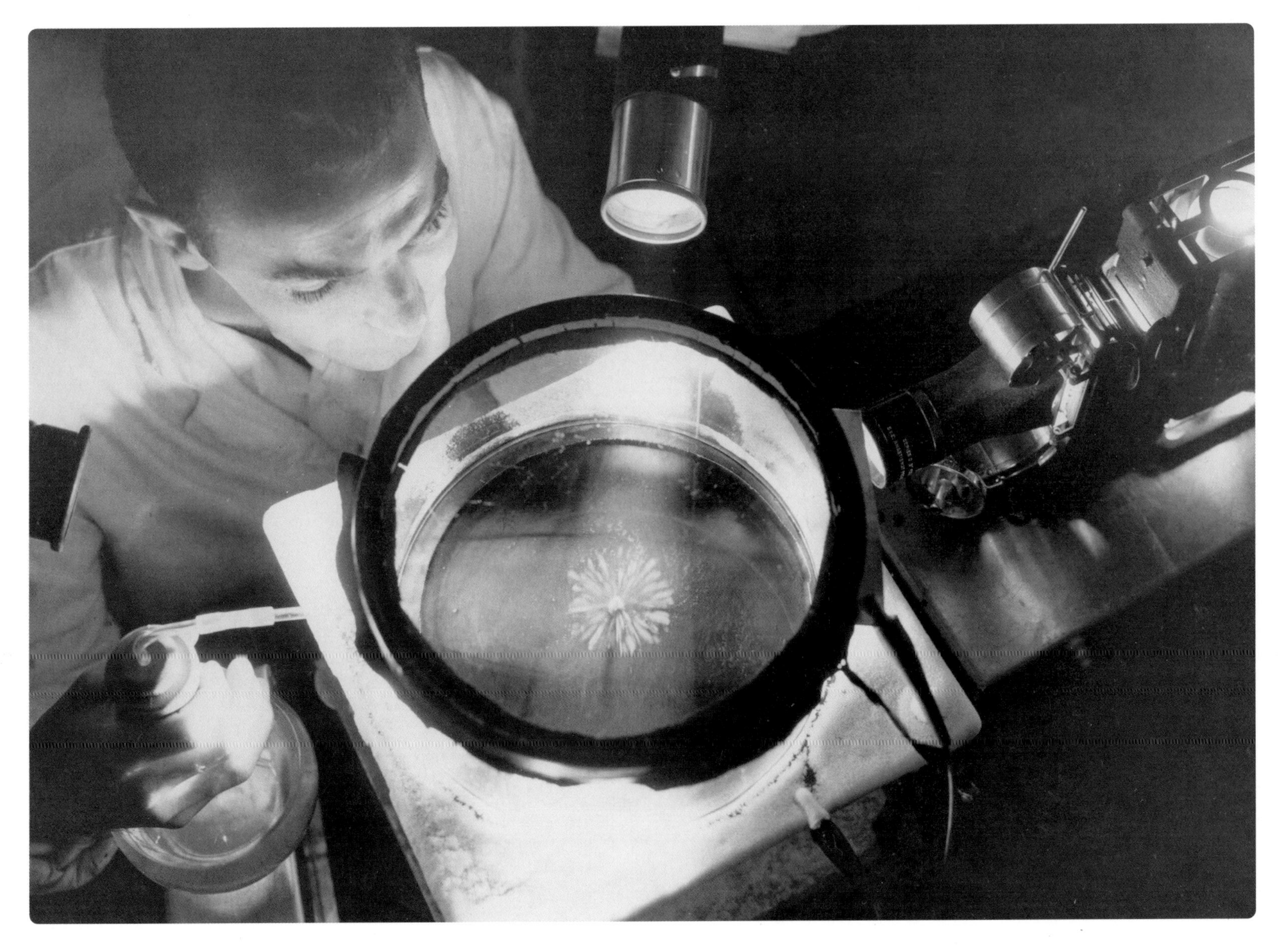

**Alpha particles emitted from a sample of polonium can be detected using a cloud chamber, where ionizing radiation causes condensation of a saturated vapor.**

# Boron

**Chemical symbol**
B

**Atomic number**
5

**Atomic mass**
10.81

**Boiling point**
7,232°F (4,000°C)

**Melting point**
3,767°F (2,075°C)

**Electron configuration**
2.3

Borax has been part of a wide variety of household cleaning and laundry products over the years, such as "Borax Extract of Soap."

**Boron is not easy to obtain as the pure element. There is not a whole lot of it in the earth's crust, and what little there is, is either tied up in compounds or tends to be contaminated with carbon.**

## BORAX

Important compounds of boron are many and various. They also have a confusing array of names. Borax, borates, and boric acid all contain boron, but are subtly different in terms of their composition and use. Borax is a particular type of borate (a compound where boron is combined with oxygen to produce a negative ion, such as $B_2O_5^{4-}$). Borax has found use as a historical cosmetic, a laundry detergent, a flux in the manufacture of steel (to lower the melting point of the alloy and to allow it to be worked more easily), a food additive, and a water softener.

Borax mines are characterized by being highly visible. The mining is "open-pit," and as such, the borax can often be seen at the surface.

## WHAT A BORE

As an element boron only came into its own in the early part of the nineteenth century. In 1808 it was first isolated almost simultaneously by the French duo of Louis Joseph Gay-Lussac (1778–1850) and Louis Jacques Thénard (1777–1857), and by English chemical giant Humphry Davy. The French favored the somewhat unfortunate moniker of bore. In the end, the name was coined by Davy from a combination of the "bor" part of borax, and the ending of the related nonmetal carbon.

## NUCLEAR APPLICATIONS

One of the fundamental tenets of producing nuclear energy is that when a nuclear fission reaction takes place, neutrons are one of the by-products. These neutrons go on to create other fission events that in turn produce even more neutrons. This sets up a chain reaction. In order to control such reactions, and to generate nuclear energy efficiently and safely, there needs to be a way of absorbing some of the neutrons that are produced and therefore controlling the reaction. Boron-10's nuclear structure (with an odd number of neutrons) makes it an excellent absorber of slow neutrons, which is why it is often used in control rods in reactors.

# Silicon

| | |
|---|---|
| **Chemical symbol** | Si |
| **Atomic number** | 14 |
| **Atomic mass** | 28.085 |
| **Boiling point** | 5,909°F (3,265°C) |
| **Melting point** | 2,577°F (1,414°C) |
| **Electron configuration** | 2.8.4 |

**Silicon is synonymous with the American high-tech industries. The connection stems from the name the element lends to the Santa Clara Valley south of San Francisco. Silicon's reach, however, extends far beyond the West Coast of the United States! As the second most abundant of all the elements, its sheer presence on earth is massive.**

### SILICON BEACH

Silicon has always been known to people, but has certainly not always been recognized as an element. The most obvious example of silicon's presence is sand, which is essentially silicon dioxide, $SiO_2$. Sand is also known as "silica."

### SILICON, SILICON EVERYWHERE

Silicon enjoys a dizzying array of applications. Sand is an essential component of glass, and silicones (compounds of silicon, oxygen, and organic carbon chains) are extensively used in applications such as silicon rubber and silicone oil lubricants, and in plastics. More recently, silicon's major impact has been in the role of a semiconductor in the electronics industry.

Perhaps the most well-known application of silicon is in the microprocessor industry, where its use as a semiconductor is ubiquitous.

# Germanium

| | |
|---|---|
| **Chemical symbol** | Ge |
| **Atomic number** | 32 |
| **Atomic mass** | 72.630 |
| **Boiling point** | 5,108°F (2,833°C) |
| **Melting point** | 1,720.85°F (938.25°C) |
| **Electron configuration** | 2.8.18.4 |

**Germanium is named after Germania, the Latin for Germany—element number 32 is one of the most historically important. Germanium has many of the properties of silicon, and much of its chemistry is related to element number 14. However, germanium is significantly less abundant than silicon in the earth's crust; in fact, of the five naturally occurring elements in group 14, it is the rarest.**

### THE KEY TO THE TABLE

Mendeleev had predicted the existence of several "as yet undiscovered" elements. One of the missing was a group 14 element that Mendeleev believed would have properties that would place it in his table between silicon and tin. He called it eka-silicon. When, in 1886, Clemens A. Winkler (1838–1904) discovered germanium in a new mineral unearthed in a German mine, the properties Mendeleev had predicted for the missing element were found to be incredibly close to the actual properties of germanium. The accuracy of the predictions about eka-silicon helped to cement the Russian's reputation.

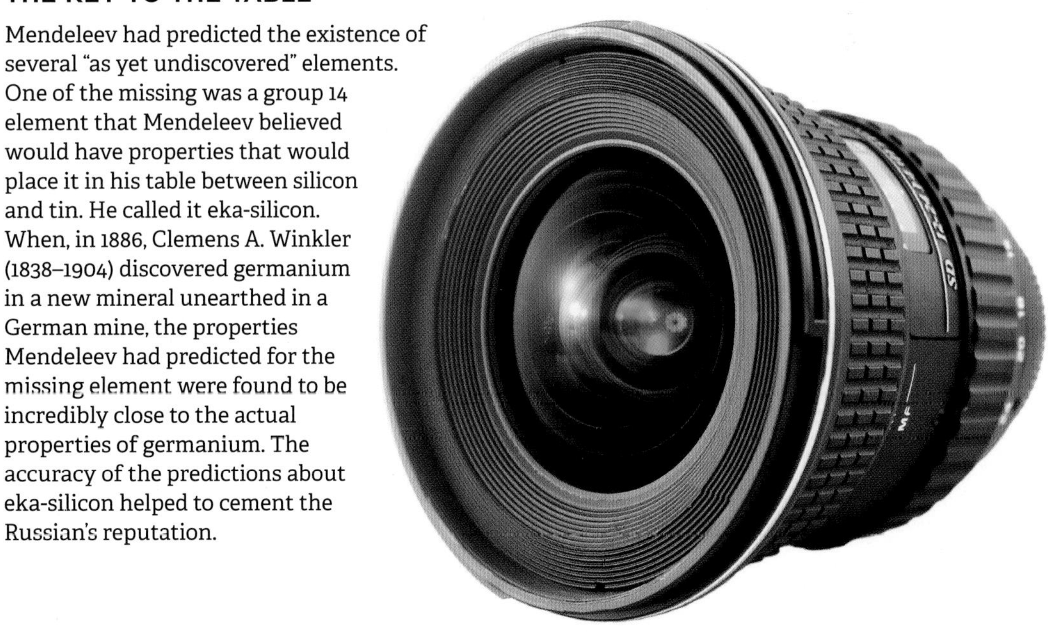

The optical properties of germanium oxide allow it to be used in wide-angle camera lenses, where its low optical dispersion makes it an ideal material.

# Arsenic

**Chemical symbol**
As

**Atomic number**
33

**Atomic mass**
74.922

**Boiling point**
1,140.8°F (sublimation 616°C)

**Melting point**
1,503°F (817°C)

**Electron configuration**
2.8.18.5

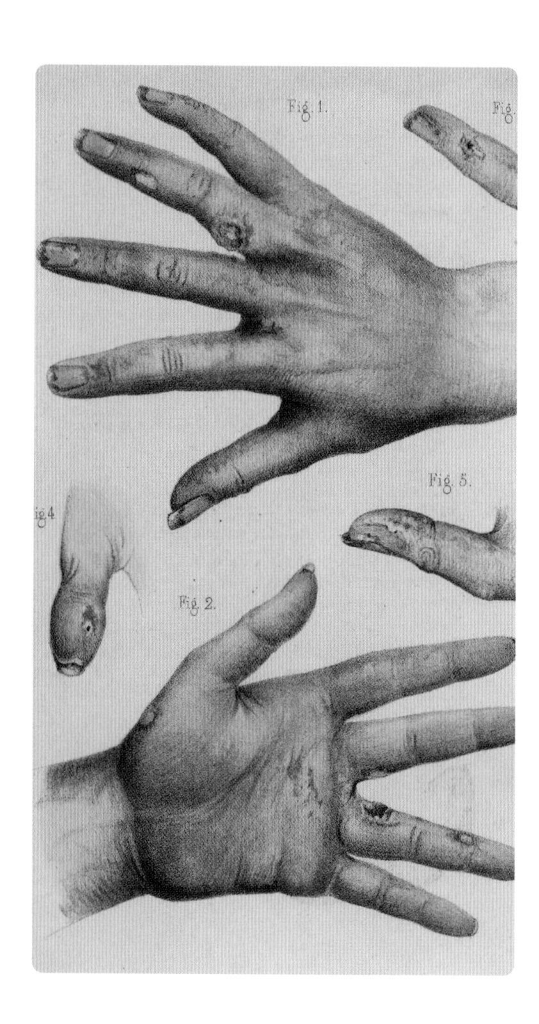

▲ This lithograph, titled "Accidents caused by the use of green arsenic dyes," was published in a French periodical in 1859.

**Arsenic is an element with an image problem. As with other elements, such as sulfur, there are many forms of arsenic, this time characterized by their colors (yellow, gray, and black), but whatever the form, arsenic tends to mean only one thing—poison!**

Arsenic is another of the ancient elements. Its discovery is sometimes attributed to Albertus Magnus (*c.* 1200–80), a thirteenth-century friar and bishop (and later a Catholic saint), but the precise history of its discovery is shrouded in doubt.

## KILL OR CURE?

Arsenic's deadly reputation is a shame, because it also has a fascinating history of use as a medicine. Most of the claims by early doctors and charlatans were pure quackery, but one Nobel Prize–winning physician, the German Paul Ehrlich (1854–1915), developed a very important, "real" medicine based on arsenic.

At the turn of the nineteenth century, the sexually transmitted disease syphilis was a serious social problem. In 1909 Ehrlich tested an array of arsenic-containing compounds on diseased rabbits, and the one that cured them came to be known commercially as Salvarsan. One had to be careful not to poison the patient, but it did work. Ultimately, penicillin proved to be a better bet.

## DYEING AND DYING

One early and popular use of arsenic compounds was in dyes and paints, in particular yellows and greens. Modern-day usage of arsenic is perhaps most important in the doping of semiconductors used in modern electronics. Here, the As atoms of group 15 replace the Si atoms of group 14 in order to increase conductivity.

Highly toxic to humans, arsenic and its compounds cause a variety of ailments—notably, serious problems with the skin and lungs. Many criminals chose arsenic as a poison in the nineteenth century because it was difficult to detect. That all changed in 1836 when the British chemist James Marsh (1794–1846) developed a test to detect minute amounts of arsenic. The test converts the arsenic oxide present in poisoning cases to arsine, $AsH_3$, and the subsequent heating of this compound creates a distinctive deposit of elemental arsenic known as a black arsenic mirror. Marsh's test was first used in the field of forensic science in France in 1840, when it was employed to solve the mariticide of Charles Lafarge.

▶ German physician Paul Ehrlich developed a cure for syphilis when he perfected the arsenic-based medicine known as Salvarsan.

# Antimony

| | |
|---|---|
| **Chemical symbol** | Sb |
| **Atomic number** | 51 |
| **Atomic mass** | 121.76 |
| **Boiling point** | 2,889°F (1,587°C) |
| **Melting point** | 11,67.13°F (630.63°C) |
| **Electron configuration** | 2.8.18.18.5 |

**Antimony's classification as a metalloid is well earned. As a metal, it is similar to lead, and throughout history it has often been mistaken for element 82, but it also has properties that are far less metallic.**

Perhaps the earliest cosmetic, kohl's use as an eyeliner goes back to ancient Egypt.

### ANTIMONY EYELINER

The ancient Egyptians and earlier civilizations knew of antimony, and one of the earliest uses of the element was as a primitive cosmetic known as kohl. In many parts of the world, tradition and belief mean the practice of applying kohl continues to this day.

### ENGINEERING METALS

Babbitt metals, invented by American goldsmith Isaac Babbitt (1799–1862), are an array of alloys of varying composition. In Babbitt metals, tin, lead, copper, and arsenic are mixed in varying percentages with antimony. The unique crystalline structure of these alloys means the materials have an extremely low coefficient of friction, so they are used as bearing surfaces where moving metal components come into contact with one another.

# Tellurium

| | |
|---|---|
| **Chemical symbol** | Te |
| **Atomic number** | 52 |
| **Atomic mass** | 127.60 |
| **Boiling point** | 1,810°F (988°C) |
| **Melting point** | 8,41.12°F (449.51°C) |
| **Electron configuration** | 2.8.18.18.6 |

**As a member of group 16 of the periodic table, tellurium shares some similarities with its nonmetal cousin sulfur, in particular its association with some wholly unpleasant odors. "Tellurium breath" is a condition in which the body converts the smallest traces of tellurium into the organic compound dimethyl telluride, $(CH_3)_2Te$, producing a foul, garlic-like odor.**

Tellurium has an unusual connection to the precious metal gold. After a lot of confusion surrounding tellurium's discovery, in 1783 Franz-Joseph Müller von Reichenstein (1740/42–1825) finally identified tellurium in a mineral he had been studying. As the chief surveyor of mining in Transylvania, Müller's job involved the analysis of unidentified ores as they came to light. At first the ore in question was thought to contain bismuth and antimony, but it was later discovered that the unusual mineral was gold telluridel. Gold is renowned for its luster. As a relatively inert element, it forms very few naturally occurring compounds, but the original ore of tellurium is one of them.

Elemental tellurium in its crystalline form takes on a silvery, metallic luster.

# Polonium

| Chemical symbol | Po |
|---|---|
| Atomic number | 84 |
| Atomic mass | 209 (longest-living isotope) |
| Boiling point | 1,764°F (962°C) |
| Melting point | 489°F (254°C) |
| Electron configuration | 2.8.18.32.18.6 |

Polonium is cited as the element that led to the death of Marie Curie's daughter, Irène Joliot-Curie (left).

Marie and Pierre Curie discovered polonium in pitchblende, the primary ore of uranium.

**One of the most politically charged elements on the periodic table, polonium has attracted more than its fair share of controversy. It was named by its discoverer, Marie Curie, after her native Poland, which was then occupied by foreign forces. Curie's hope was that her homeland's plight might be brought to worldwide attention, and that independence might follow.**

## THE CURIES

Polonium is found in uranium ores, but it is a very rare element and was therefore difficult to find and isolate. Marie Curie and her husband, Pierre, extracted the first sample of the element from pitchblende, a uranium-rich ore, in 1898. Their interest was aroused as the pitchblende that they were studying was emitting an enormous amount of radioactivity, which could not be attributed to the uranium alone. Polonium is brutally radioactive as an alpha particle emitter, and is extremely dangerous. Sadly, it is cited as the element that caused Marie Curie's daughter Irène Joliot-Curie's (1897–1956) death from leukemia in 1956, following her exposure to a broken vial of the element ten years earlier.

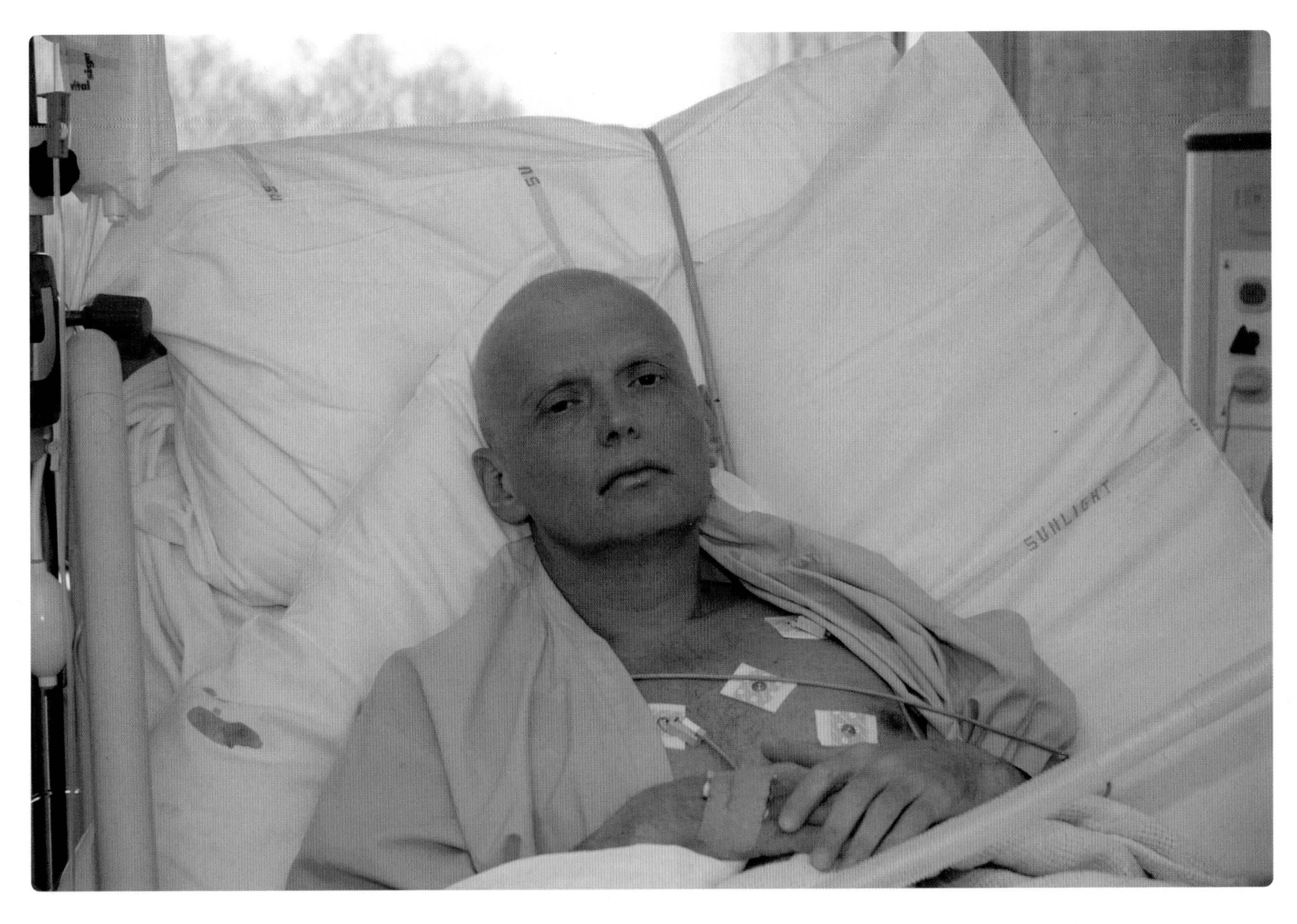


Russian dissident Alexander Litvinenko in his hospital bed in the Intensive Care Unit of University College Hospital, London. He died in 2006 as a result of being poisoned with radioactive polonium-210.

## POLONIUM-210

Polonium is a soft, silvery-gray element that forms a large number of isotopes, but by far the most infamous is polonium-210.

When Alexander Litvinenko fled to London in 2000, the former Russian secret service officer was already in hot water. He had publicly accused the state of ordering assassinations for political gain, and he had been arrested and incarcerated in his native country. When he began to work with the British intelligence services, and then wrote two books with further accusations against the Russian hierarchy, his fate was sealed. After eating out at a London restaurant in November 2006, Litvinenko suddenly fell ill and was hospitalized. Three weeks later he was dead. Litvinenko had been exposed to an unusually large dose of polonium-210, and he died as a result of radiation sickness caused by the deadly isotope.

# LANTHANOIDS

**As in many other examples of our grouping of elements, the lanthanoids have their own complicated history of elemental inclusion and exclusion, with, in this case, the addition of a terribly misleading collective name. Sometimes referred to as the rare earth elements, that moniker creates problems of its own that only serve to muddy the waters further.**

On the following pages:

| | | | | | |
|---|---|---|---|---|---|
| **La** | Lanthanum | **Sm** | Samarium | **Ho** | Holmium |
| **Ce** | Cerium | **Eu** | Europium | **Er** | Erbium |
| **Pr** | Praseodymium | **Gd** | Gadolinium | **Tm** | Thulium |
| **Nd** | Neodymium | **Tb** | Terbium | **Yb** | Ytterbium |
| **Pm** | Promethium | **Dy** | Dysprosium | **Lu** | Lutetium |

| 1 | 2 | 3 | 4 | 5 | 6 | 7 | 8 | 9 | 10 | 11 | 12 | 13 | 14 | 15 | 16 | 17 | 18 |
|---|---|---|---|---|---|---|---|---|---|---|---|---|---|---|---|---|---|
| 1 H | | | | | | | | | | | | | | | | | 2 He |
| 3 Li | 4 Be | | | | | | | | | | | 5 B | 6 C | 7 N | 8 O | 9 F | 10 Ne |
| 11 Na | 12 Mg | | | | | | | | | | | 13 Al | 14 Si | 15 P | 16 S | 17 Cl | 18 Ar |
| 19 K | 20 Ca | 21 Sc | 22 Ti | 23 V | 24 Cr | 25 Mn | 26 Fe | 27 Co | 28 Ni | 29 Cu | 30 Zn | 31 Ga | 32 Ge | 33 As | 34 Se | 35 Br | 36 Kr |
| 37 Rb | 38 Sr | 39 Y | 40 Zr | 41 Nb | 42 Mo | 43 Tc | 44 Ru | 45 Rh | 46 Pd | 47 Ag | 48 Cd | 49 In | 50 Sn | 51 Sb | 52 Te | 53 I | 54 Xe |
| 55 Cs | 56 Ba | **57-71** | 72 Hf | 73 Ta | 74 W | 75 Re | 76 Os | 77 Ir | 78 Pt | 79 Au | 80 Hg | 81 Tl | 82 Pb | 83 Bi | 84 Po | 85 At | 86 Rn |
| 87 Fr | 88 Ra | 89-103 | 104 Rf | 105 Db | 106 Sg | 107 Bh | 108 Hs | 109 Mt | 110 Ds | 111 Rg | 112 Cn | 113 Nh | 114 Fl | 115 Mc | 116 Lv | 117 Ts | 118 Og |
| | | | **57 La** | **58 Ce** | **59 Pr** | **60 Nd** | **61 Pm** | **62 Sm** | **63 Eu** | **64 Gd** | **65 Tb** | **66 Dy** | **67 Ho** | **68 Er** | **69 Tm** | **70 Yb** | **71 Lu** |
| | | | 89 Ac | 90 Th | 91 Pa | 92 U | 93 Np | 94 Pu | 95 Am | 96 Cm | 97 Bk | 98 Cf | 99 Es | 100 Fm | 101 Md | 102 No | 103 Lr |

First, considering the term "rare earth" is used to describe them, it would seem reasonable to think that these elements are, well, rare. Not so. In fact, some of the elements are relatively abundant in the earth's crust compared to many others found on the periodic table. Second, because of their extreme chemical similarity (much of their chemistry is based upon the +3 oxidation state indicated by the generic $Ln^{3+}$), they proved to be an extraordinarily difficult group of elements to separate from one another and to identify as distinct. Their history is littered with mistaken identity, false claims, and general confusion.

## SUPPLY AND DEMAND

The rare earths/lanthanoids have been thrown into the spotlight in recent years because of the astonishing growth in the personal electronics industry. Many of the elements in this collection are used in the production of cell phones, laptops, and all manner of other small electronic devices; as such, their sustainability is of increasing concern. Matters are further complicated by geopolitics, with China playing a major role as both a supplier and a user of lanthanoids. One solution is to seek alternative materials, but not all of the elements can be easily substituted.

**The ubiquitous presence of so many lanthanoid metals in a vast array of personal electronics, such as cell phones and laptops, means that they are valuable resources, but with demand increasing, their sustainability is under question.**

# Lanthanum

| | |
|---|---|
| **Chemical symbol** | La |
| **Atomic number** | 57 |
| **Atomic mass** | 138.91 |
| **Boiling point** | 6,267°F (3,464°C) |
| **Melting point** | 1,688°F (918°C) |
| **Electron configuration** | 2.8.18.18.9.2 |

**Lanthanum is the first, the lightest, and one of the more abundant of the lanthanoids, and of course, the element the series is named for. The lanthanoids are a nightmare to get at because not only do they cluster together in their ores, but their chemistry is similar.**

The Swede Carl Gustaf Mosander was a prolific chemist, not only discovering lanthanum, but also erbium and terbium among the lanthanoids.

When Carl Gustaf Mosander (1797–1858) discovered lanthanum in a sample of cerium nitrate in 1839, he turned to the problems associated with isolating the rare earth elements from one another as inspiration for element number 57's name. He named it after the Greek *lanthanein*, meaning "to lie hidden."

### DATING ROCKS

An isotope of lanthanum, La-138, can be utilized in the radioisotope dating of very old rocks. La-138 has a half-life of $1.05 \times 10^{11}$ years, and one of its decay products is an isotope of barium, Ba-138. By measuring the relative amounts of these isotopes present in a lump of ancient rock, geologists can make decent estimates of the age of some extremely old geological forms.

# Cerium

| | |
|---|---|
| **Chemical symbol** | Ce |
| **Atomic number** | 58 |
| **Atomic mass** | 140.12 |
| **Boiling point** | 6,229°F (3,443°C) |
| **Melting point** | 1,470.2°F (799°C) |
| **Electron configuration** | 2.8.18.19.9.2 |

**Cerium is the most abundant of all the rare earth elements, and in fact, as the twenty-fifth most abundant of all the elements in the earth's crust, it is actually anything but "rare."**

In the first few years of the nineteenth century, interest in celestial bodies was aroused by the observation in 1801 of a new asteroid, named Ceres. A couple of years later, a new chemical element was found. In 1803, Wilhelm Hisinger (1766–1852)—with the help of Jöns Jacob Berzelius—discovered element number 58, and the new lanthanoid was named after the recently observed asteroid.

### BURN, BABY, BURN

Cerium's ability to produce sparks means that it is used as a flint in misch metals, and it can be combined (chiefly with iron) in an almost identical material called, appropriately, ferrocerium. Ferrocerium has been marketed under a number of different trade names as a fire-starter for camping enthusiasts.

### REPLACING RED

Another compound of cerium, this time an oxide, turns up in a couple of places. It is insoluble so can be added to water and then applied to glass, where it acts as a fine abrasive, polishing the glass in specialty lenses.

Ferrocerium produces sparks when struck against a rough material, making it a useful material for fire-starters and cigarette lighters.

# Praseodymium

**Chemical symbol**
Pr

**Atomic number**
59

**Atomic mass**
140.91

**Boiling point**
6,368°F (3,520°C)

**Melting point**
1,707.8°F (931°C)

**Electron configuration**
2.8.18.21.8.2

**It's next to impossible to separate praseodymium and its periodic table neighbor and twin element, neodymium, from one another. Praseodymium is taken from the Greek *prasinos didymos*, meaning "green twin," and neodymium from *neos didymos*, meaning "new twin."**

Mosander—the discoverer of lanthanum—thought he had discovered a new element mixed up in samples of the already known cerium and lanthanum. He called this new "element" "didymium," after the Greek word for "twin." Even though it turned out that didymium was not an element at all, Mosander's choice of name could hardly have been more prophetic.

## SEPARATING THE TWINS

It was not until spectroscopy came into its own that there was a decent chance of identifying many of the lanthanoids in any definitive way. Enter Carl Auer von Welsbach (1858–1929), a student of Robert Bunsen's, and hence well positioned to exploit his spectroscopy technique to aid with further element identifications. He also went on to confirm that didymium was in fact two elements—Mosander's original name could hardly have been a better one, but for a different reason than he had originally imagined! Named for the color of its oxide, praseodymium (along with its "twin," neodymium) was discovered in 1885.

## BACK TOGETHER IN GLASSES

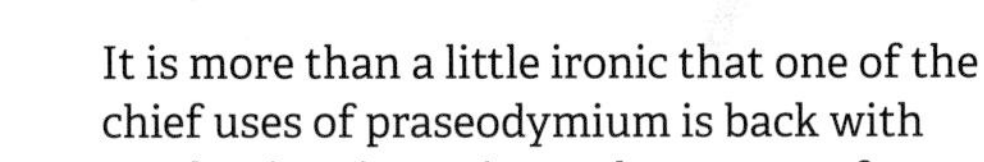

It is more than a little ironic that one of the chief uses of praseodymium is back with neodymium in a mixture known as, of course, didymium—Mosander would have been delighted! Mixed together, the two lanthanoid elements are used to make the lenses of safety spectacles used by glassblowers. When mixed, the metals have a peculiar property. Their combination leads to a substance that allows the intense orange-yellow light that is emitted when glass that contains sodium is heated to be completely filtered out, thus protecting the craftsman's eyes from damage.

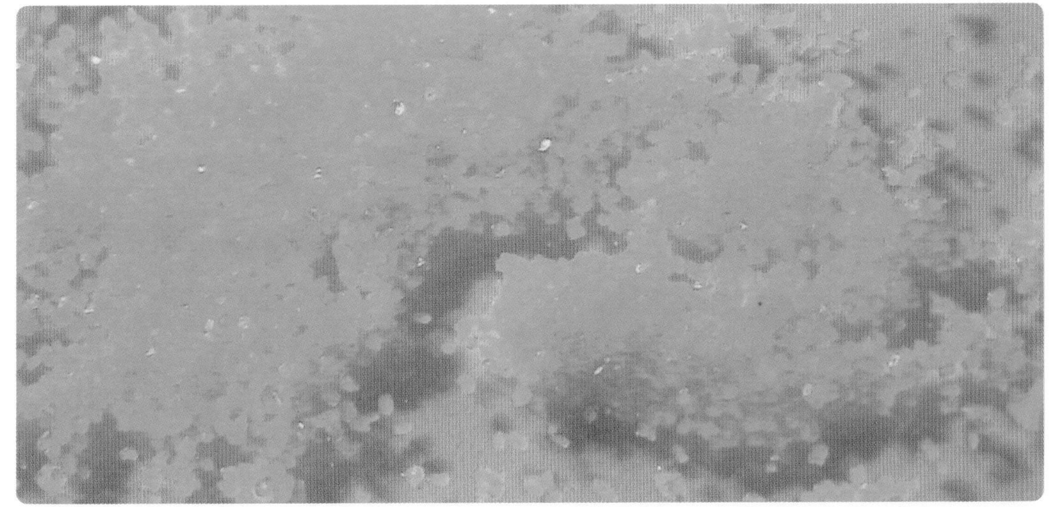

A hydrate sample of praseodymium(III) sulfate with eight water molecules associated with each formula unit of the salt.

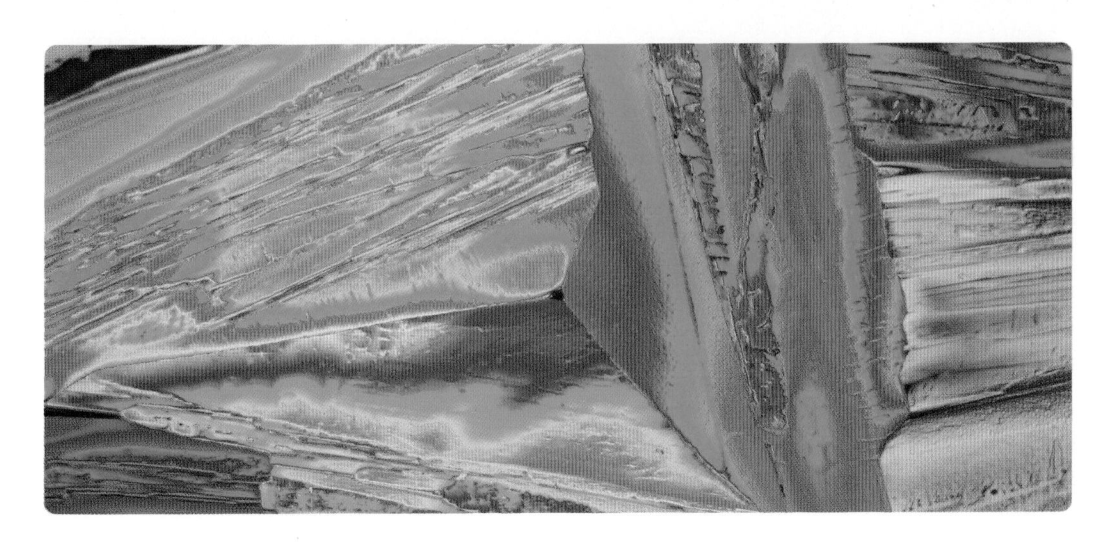

A sample of an aqueous solution of praesodymium nitrate, viewed on a microscopic slide, and illuminated with polarized light.

# Neodymium

**Chemical symbol**
Nd

**Atomic number**
60

**Atomic mass**
144.24

**Boiling point**
5,565°F (3,074°C)

**Melting point**
1,869.8°F (1,021°C)

**Electron configuration**
2.8.18.22.8.2

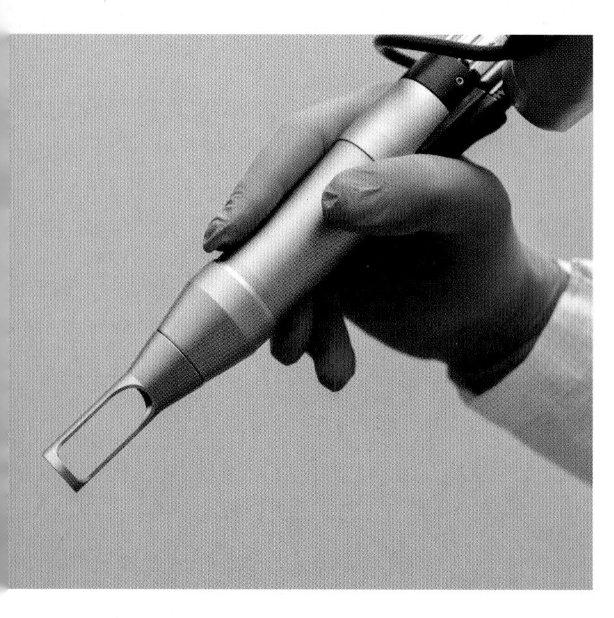

YAG lasers, whch are doped with neodymium, are used in medical and dental applications.

**Despite the fact that, in many senses, elements 59 and 60 are almost totally entwined, neodymium has one reason to stand out on its own. Thanks to one single property, first discovered in the 1970s, neodymium might be the most well-known and most familiar of all of the lanthanoids.**

## MAGNETIC MADNESS

The combination of neodymium, iron, and boron makes for some ferociously strong, permanent magnets. When combined in the correct ratio, $Nd_2Fe_{14}B$, the elements together create an ultra-strong magnetic field. This is partly due to the crystal structure of the compound formed, and partly due to the fact that neodymium has seven unpaired electrons in its electron configuration. Unpaired electrons create their own tiny magnetic fields, and when billions of atoms are lined up in a single direction, the accumulation of these tiny fields creates a very strong magnetic force. Recent research has suggested that neodymium magnets can be used to open magnetically sensitive ion channels in the brain, resulting in the noninvasive tweaking of neurons, and hence potentially helping with the treatment of neurological disorders.

## DO NOT EAT!

A common accident involves a child swallowing a pair of magnets. Once ingested, the super-strong magnets have a habit of rapidly finding one another, and violently snapping together. If that occurs when one magnet is in one part of the intestine or stomach, and the other magnet is in another part, then as the magnets collide they can pinch and perforate the small intestine, bowel, or stomach wall.

## LASER LIGHT

Neodymium has an important application in didymium glass, and it has a few other, quite specialized applications too. One is in the production of lasers. Here, neodymium is used as a doping agent, where it introduces a small impurity into a crystal that is then used to produce the laser. These lasers have several applications, including in medicine, where they have been used for corrective eye surgery.

Neodymium magnets are the strongest permanent magnets. They have many applications, including in computer disk drives and MRI machines.

# Promethium

**Chemical symbol** Pm

**Atomic number** 61

**Atomic mass** 145 (longest-living isotope)

**Boiling point** 5,432°F (3,000°C)

**Melting point** 1,908°F (1,042°C)

**Electron configuration** 2.8.18.23.8.2

**Promethium suffers from the same issues as technetium inasmuch as it has no stable isotopes and so is basically "missing" from the earth's crust. Add to the mix the difficulty of separating one rare earth from another, and as a result, element number 61 was a tough one to find. It was 1945 before its discovery was finally confirmed.**

## FALSE STARTS

Before it was named after the Greek Titan Prometheus, promethium had at least two other names. In the mid-1920s two serious claims were made on element 61. Scientists at the University of Illinois thought that they had observed spectral lines that could be attributed to the missing lanthanoid. They christened the element "illinium," after their home state. Two Italian scientists working at the Royal University in Florence conducted similar research, and named their version of the element "florentium." Both claims were ultimately dismissed and credit finally went to the Oak Ridge National Laboratory for detecting promethium in 1945.

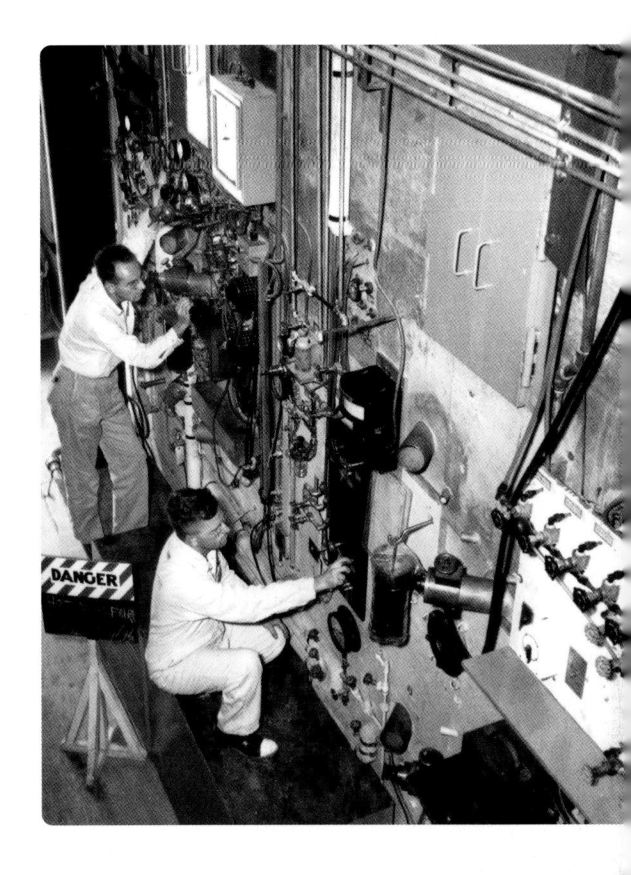


The Oak Ridge National Laboratory graphite reactor in Oak Ridge, Tennessee, where element number 61 was first detected, in 1945.

# Samarium

**Chemical symbol** Sm

**Atomic number** 62

**Atomic mass** 150.36

**Boiling point** 3,261.2°F (1,794°C)

**Melting point** 1,965.2°F (1,074°C)

**Electron configuration** 2.8.18.24.8.2

**Element number 62 could be described as the first element to be named after a living person. A shiny metal that reacts with oxygen, it is the second most abundant of the lanthanoids, being made up of a number of naturally occurring isotopes. Three of those isotopes have extraordinarily long half-lives, which has led to their use in a complicated dating technique called geochronology.**

Samarium can be found in a number of rare earth ores, notably cerite, gadolinite, samarskite, monazite, and bastnäsite. One of its isotopes, Sm-147, has an extraordinarily long half-life of $1.06 \times 10^{11}$ years, and this is useful in dating rocks and other geological formations.

## EPONYMOUS ELEMENTS?

Samarium was actually named after the mineral that it was extracted from, samarskite. However, the German mineralogist Heinrich Rose (1795–1864) had, in turn, named samarskite after Vasili Samarsky-Bykhovets (1803–1870). Samarsky-Bykhovets was a high-ranking Russian mine official who provided the original sample of the mineral for analysis.

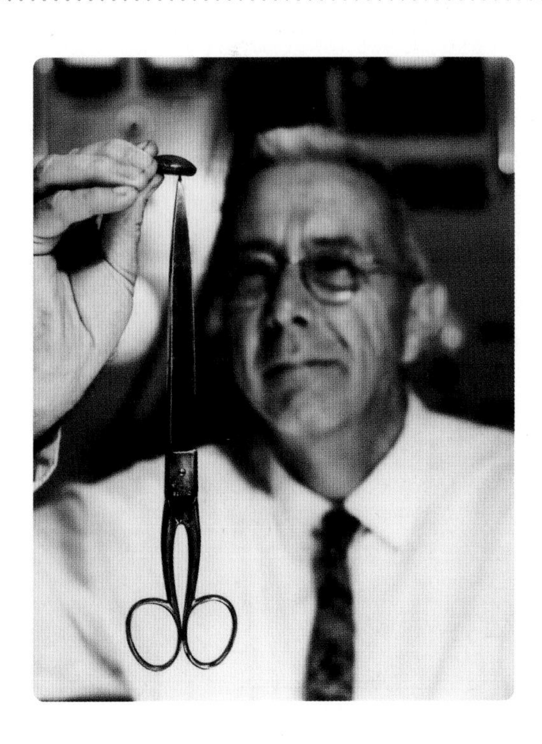

A 1968 demonstration of the power of a magnet made from samarium and cobalt. It was then one of the most powerful small magnets ever made.

# Europium

**Chemical symbol**
Eu

**Atomic number**
63

**Atomic mass**
151.96

**Boiling point**
2,784°F (1,529°C)

**Melting point**
1,511.6°F (822°C)

**Electron configuration**
2.8.18.25.8.2

**Europium started life as an offshoot of its periodic table neighbor samarium. Europium had a few of the early spectroscopy gurus scratching their heads, with William Crookes and Paul-Émile Lecoq de Boisbaudran both noting some oddities in samarium samples. But it was left to Frenchman Eugène-Anatole Demarçay (1852–1903) to confirm europium, and isolate one of its salts in 1901. Why he called it europium, nobody seems quite sure.**

Demarçay became an expert in spectroscopy, analyzing experimental data to find new elements (in the case of europium), and helping in the confirmation of others like the Curies' radium.

▲ The use of inks containing europium has greatly helped in the fight against the counterfeiting of currency.

## RED AND BLUE

Europium first found utility through the phosphorescence of its compounds, notably as $Eu^{3+}$ and $Eu^{2+}$. The latter is an example of a lanthanoid element finding favor in an oxidation state other than +3, a relatively unusual occurrence. Red colors are produced from phosphors that contain compounds with the higher-charged $Eu^{3+}$ ion, such as $Eu_2O_3$, and europium in the 3+ state was used in old-fashioned televisions. Compounds of the lower oxidation state (+2) emit blue light and they are sometimes combined with the red europium(III) compounds and fellow lanthanoid terbium(III) compounds (green), in order to produce manufactured white light.

▲ When the pure metal is exposed, a yellow film forms on the surface of europium.

## TAKE NOTE, NO FORGERY

Continuing europium's utility as an optical element, we find it in inks that are used to print paper currency. When viewing such currency under ultraviolet (UV) light, the europium ions absorb the UV light, and the electron movement caused by the transfer of such electromagnetic energy is responsible for the red light that is emitted. By incorporating the exotic inks into the notes, they become increasingly tricky to counterfeit convincingly. Another form of "currency," the humble postage stamp, has also benefited from the chemistry of element number 63. Europium oxide has been incorporated into stamps to allow machines to recognize the glow given off by the europium ions, in order to establish the stamps' face value.

## RAYS AND RODS

A number of patents have been filed for europium-doped polyethylene plastics, which have been shown to help enhance plant growth. These sheets, once again through the optical properties of europium, convert UV light present in natural sunlight to a light source that the plants can use in the photosynthesis process. Another good neutron absorber, europium also appears in some control rods for nuclear reactors.

# Gadolinium

| Chemical symbol | Gd |
|---|---|
| Atomic number | 64 |
| Atomic mass | 157.25 |
| Boiling point | 5,923.4°F (3,273°C) |
| Melting point | 2,395.4°F (1,313°C) |
| Electron configuration | 2.8.18.25.9.2 |

**Gadolinium has a special place in the history of the lanthanides, as revealed by its name. Ask a chemist who the most important figure in the history of the lanthanoids is, and they may well answer with one name: Finnish chemist Johan Gadolin. Element number 64 is named for Gadolin, an appropriate accolade for a man who has been called "the father of the rare earths."**

Much of the chemistry of the lanthanoids revolves around their +3 oxidation state, and it is in this state that gadolinium finds a major medical application as a contrasting agent in MRI. A contrasting agent helps to maintain the quality of the image, and therefore aids diagnosis, but is potentially very toxic.

Many health issues can arise when one element in its ionic form mimics another, possibly replacing it or disrupting that naturally occurring ion's proper function. When one learns that $Gd^{3+}$ has a similar size to $Ca^{2+}$, a ubiquitous ion in the body, one can see how many biological functions could be disrupted.

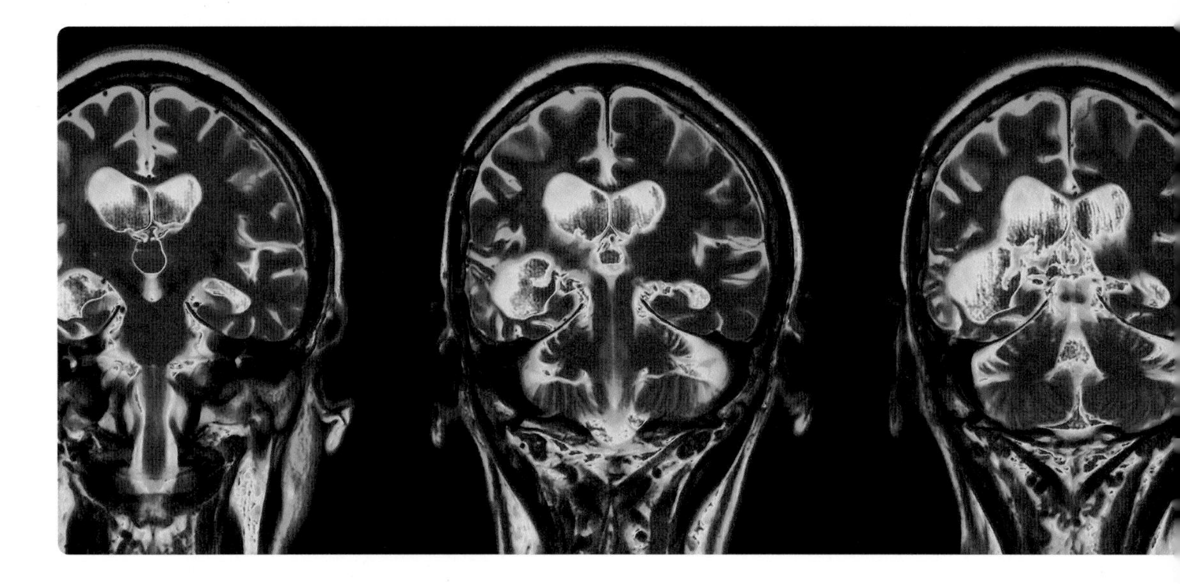

MRI uses gadolinium, but element number 64 may be on the way out in this application, as concerns over its toxicity continue to mount.

# Terbium

| Chemical symbol | Tb |
|---|---|
| Atomic number | 65 |
| Atomic mass | 158.93 |
| Boiling point | 5,846°F (3,230°C) |
| Melting point | 2,473°F (1,356°C) |
| Electron configuration | 2.8.18.27.8.2 |

**The chemistry of the lanthanoids, in particular their extraction from single sources and the nightmarish difficulties of separating them all out, makes things complicated enough, but what about when elements that have almost identical names get accidentally switched around? Well, it makes for even deeper confusion, and that is exactly what happened with terbium and its sister element number 68, erbium.**

In keeping with the procedure of continually fine-tuning the oxides of the lanthanoids, the early 1840s saw Mosander extract two more elements from an original sample of the oxide of yttrium. The year was 1843, and the oxides that Mosander pulled from the yttria were distinct from one another in terms of their colors. The first one was yellow, and he called this erbium oxide (erbia). The second one was pink, and he called it terbium oxide (terbia). Each name was derived from the Ytterby name, and because they were found at the same time and from that same source, it all seemed to make sense.

## MIX AND MATCH

Sometime in the early 1860s, Swiss chemist Marc Delafontaine managed to mix up Mosander's erbia and terbia. Delafontaine made the old erbia the new terbia, and vice versa. This mistake persists to this day, so when one searches online for an image of erbium(III) oxide, one will find a beautiful array of pink powders and crystals, and not Mosander's original yellows.

Terbium is a silvery white, relatively soft metal that is ductile and malleable.

# Dysprosium

**Chemical symbol**
Dy

**Atomic number**
66

**Atomic mass**
162.5

**Boiling point**
4,653°F (2,567°C)

**Melting point**
2,573.6°F (1,412°C)

**Electron configuration**
2.8.18.28.8.2

**The Greek word from which the name of dysprosium is derived coud equally have been applied to any one of the fifteen lanthanoids. *Dysprositos*, meaning "hard to get at," was so named by its discoverer, Paul-Émile Lecoq de Boisbaudran, at least in part for the methodology he had to employ to get to and identify dysprosium.**

## PRECIPITATING OVER AND OVER AND OVER

Like so many of the other lanthanoids, dysprosium was discovered as a result of the continued refinement of a sample of something that was already thought to be pure, this time a sample of holmium. In this respect, the rare earths are somewhat like one of those Russian *matryoshka* dolls, where one just keeps popping out from inside another. The story goes that in the late 1800s Lecoq de Boisbaudran had repeatedly to put the sample in solution, and then perform more than fifty precipitations using various acids, before he got to a sample of a compound of element number 66. A sample of the pure element was not produced until the early 1950s, via the ion-exchange method developed by Frank Spedding (1902–84) at the University of Iowa.

Frank Spedding (right) with the physicist Niels Bohr. Spedding was a pioneer in the separation of the lanthanoids.

## SPEDDING SPEEDS THE RESEARCH

Because it had been so difficult to obtain decent samples of the lanthanoid elements, there was unlikely to be any research into their chemistry, and therefore their potential uses, until a practical method for processing and separating the ores was developed. As an example of Frank Spedding's important work, one 1950 paper he co-authored was entitled "Separation of Rare Earths by Ion Exchange. IV. Further Investigations Concerning Variables Involved in the Separation of Samarium, Neodymium, and Praseodymium." Spedding and his team's efforts were so vital in extracting the lanthanoids, and therefore in finding uses for them, that his impact cannot be underestimated.

## GOT IT. NOW WHAT?

So much for actually getting to dysprosium; but what about its uses? Whether it is as doping material in specialized ceramics, or as small parts of various magnetic materials, or as a neutron absorber, like the other lanthanoids, all of element 66's applications are quite specialized. One of the more mainstream uses of dysprosium has been in the form of the triiodide and tribromide compounds, $DyI_3$ and $DyBr_3$, respectively, in metal-halide lighting.

Dysprosium is used in metal-halide lighting. The intensity and color of the light can be controlled by varying the composition of the compounds and gases present.

# Holmium

| | |
|---|---|
| **Chemical symbol** | Ho |
| **Atomic number** | 67 |
| **Atomic mass** | 164.93 |
| **Boiling point** | 4,892°F (2,700°C) |
| **Melting point** | 2,685.2°F (1,474°C) |
| **Electron configuration** | 2.8.18.29.8.2 |

**Although it is not one of the four elements that was named for that famous town of Ytterby in Sweden, holmium nevertheless has its roots firmly ensconced in Scandinavia. As the infernally complicated ores of the lanthanoids were purified, holmium was a result of finer and finer separation.**

Most sources give joint credit for the discovery of holmium to Marc Delafontaine (1837–1911), Jacques-Louis Soret (1827–1890), and Per Teodor Cleve (1840–1905) in 1878. Holmium came from erbium samples, which in turn contained yttrium. Later still, holmium samples yielded dysprosium. One could say that Cleve really won the day, because, like hafnium, the name for element 67 is derived from the Latin name for a Scandinavian city, this time Holmia, or Stockholm.

## MORE MAGNETS, MORE IMAGES

Holmium has a couple of close connections to one of its near neighbors, neodymium. First, holmium has curious magnetic properties. It is the lanthanide that exhibits the strongest paramagnetism, meaning that when it is placed in an external magnetic field, it has the ability to concentrate the magnetic field to the greatest degree. This superpower is put to use in MRI machines, which allow detailed pictures of the inside of the body to be taken.

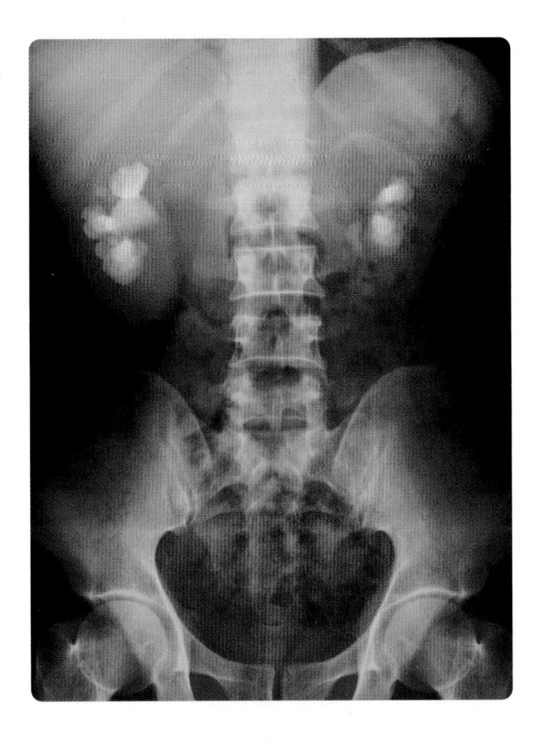

▲ Although holmium is one of the least utilized elements, it finds important applications in medicine, in both laser and imaging technology.

# Erbium

| | |
|---|---|
| **Chemical symbol** | Er |
| **Atomic number** | 68 |
| **Atomic mass** | 167.26 |
| **Boiling point** | 5,194°F (2,868°C) |
| **Melting point** | 2,784°F (1,529°C) |
| **Electron configuration** | 2.8.18.30.8.2 |

**Having been extracted and confused with terbium right from the beginning, erbium (or more accurately erbia) itself became the source of yet another of the lanthanoids. After Carl Gustaf Mosander had finished with his erbium oxide, it was left for Jean Charles Galissard de Marignac to pull ytterbium from it.**

## PINK AND GREEN

Erbium's pale pink salts are stunning, and among chemists color is probably erbium's best loved attribute. White light is a mixture of all of the colors of the rainbow, the classic ROYGBIV (red, orange, yellow, green, blue, indigo, and violet), and it's that fact that accounts for the color of all objects. For any compound to exhibit color, it must absorb certain wavelengths of light and reflect others. Erbium happens to absorb light in the green part of the spectrum, with a wavelength of around 530 nanometers. As the green light is absorbed by the erbium ions, what's left is reflected. Since pink is a complementary color to green, pink light is reflected.

## EYEWEAR AND SAFETY

As well as absorbing light in the green part of the spectrum, erbium also has the ability to absorb infrared light. This is particularly useful if one wants to extract infrared light from an incoming source to protect one's eyes, and as a result, erbium-impregnated safety eyewear has been developed.

▲ Erbium oxide, $Er_2O_3$, has a very distinctive pink color. It has a number of optical and electrical applications.

# Thulium

| | |
|---|---|
| **Chemical symbol** | Tm |
| **Atomic number** | 69 |
| **Atomic mass** | 168.93 |
| **Boiling point** | 3,542°F (1,950°C) |
| **Melting point** | 2,813°F (1,545°C) |
| **Electron configuration** | 2.8.18.31.8.2 |

**Another element that came from an impure erbium oxide, thulium was discovered by Per Teodor Cleve in 1879. Cleve was a Swede, and not content with paying homage to his native country by naming element number 67 after its capital city, he named number 69 thulium, after Thule, an ancient name for the northernmost region of Europe.**

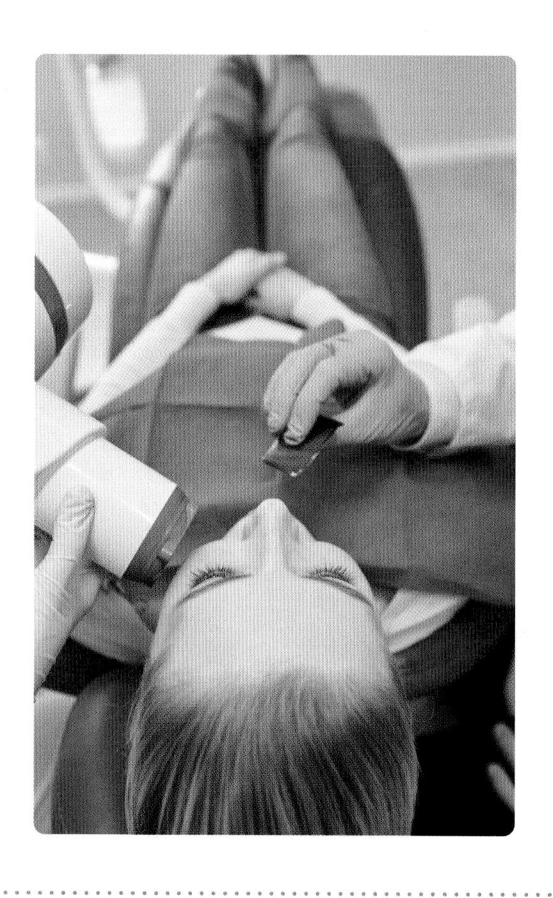

### REPLACING GD?

In recent years there has been increased concern over the potential for gadolinium ions used in MRI machines to escape their ligand cages and enter the body. As a result, scientists have been seeking alternatives to gadolinium, and thulium is one such potential replacement.

### X-RAY SOURCE

Thulium was proposed as a source of X-rays as early as the mid-1950s. Tm-170 is the isotope in question. It was originally produced in a heavy-water nuclear reactor. Thulium is an ideal material for low-intensity, low-tech, simple, and portable X-ray machines. With a half-life of just under 129 days, a small piece of thulium can remain useful as an X-ray source for a couple of years before it needs to be replaced.

Thulium has found use in low-intensity, portable X-ray machines, such as those found in dentists' offices.

# Ytterbium

| | |
|---|---|
| **Chemical symbol** | Yb |
| **Atomic number** | 70 |
| **Atomic mass** | 173.05 |
| **Boiling point** | 2,185°F (1,196°C) |
| **Melting point** | 1,506.2°F (819°C) |
| **Electron configuration** | 2.8.18.32.8.2 |

**If you think that an element with such an odd-looking name simply must be related to the other element on the periodic table with an equally odd-looking one, then you'd be right. As the final of the four elements to be named after the Swedish town of Ytterby, element number 70 forever has that connection to element number 39, yttrium (as well as to terbium and erbium).**

In fact, impure samples of yttrium gave up erbium, number 68, before samples of it gave up ytterbium. When, in 1878, the Swiss chemist Jean Charles Galissard de Marignac isolated the new oxide, he named the element in such a way that it paid homage to both yttrium and erbium—entirely appropriately.

### ALL CHANGE

Bright, shiny, ductile, and malleable, ytterbium is a fairly standard-looking metal. It has a variety of naturally occurring isotopes with masses ranging from 168 to 176, with ytterbium-174 being the most abundant. Those isotopes, taken together with their relative natural abundances considered, used to give an average atomic mass for ytterbium of 173.054. That was until, in 2015, that mass was revised by the IUPAC to 173.045, when a reassessment of the isotopic abundance of the element was incorporated into the new calculation.

Ytterbium finds a special application in atomic clocks, where it is set to improve upon even the ultra-accurate cesium versions.

# Lutetium

**Chemical symbol**
Lu

**Atomic number**
71

**Atomic mass**
174.97

**Boiling point**
6,156°F (3,402°C)

**Melting point**
3,025.4°F (1,663°C)

**Electron configuration**
2.8.18.32.9.2

**There are several elements that have a French connection. A couple of them are fairly obvious, with gallium and francium named after the Latin and the modern name for France, respectively. Then there's the whole Marie Curie link. Element number 71 is one of these elements too, but you'd have to be a scholar of language to know that the Latin name for Paris was Lutetia.**

Lutetium didn't arrive until 1907, making it the last of the lanthanoids to be discovered, and the ultimate credit goes to Georges Urbain (1872–1938) who was, you've guessed it, a native of Paris. The discovery was not concluded until after Charles James (1880–1928) in the United States had isolated a large amount of the element—but never published his findings—and Carl Auer von Welsbach had done something similar in Germany. Auer von Welsbach was a little more audacious than the English-born James, and went as far as to name his element. He called it cassiopeium—another celestial reference, this time to a constellation—and this name persisted in the literature, especially in Germany, through the 1950s.

## WHERE IS LU?

Lutetium's positioning on the periodic table remains open to question. If you look at the current IUPAC periodic table, you will see it where you are probably expecting it, at the far right of the 4f block that has been separated out before the main body of the table. However, lutetium is sometimes placed in group 3 of the periodic table, underneath yttrium, element number 39, and to the left of hafnium, element number 72. Lutetium is a hard, silvery-white metal and its ions are the smallest of the $Ln^{3+}$ ions, a result of what is known as the lanthanide contraction. The size of the $Lu^{3+}$ ion makes it quite like that of $Sc^{3+}$ and $Y^{3+}$.

## RARE AND EXPENSIVE

Lutetium is the rarest of the lanthanoids and that makes it very expensive. However, this is offset by the fact that lutetium has very few uses and therefore is not in especially high demand. Its use as a catalyst component in petroleum cracking and other industrial processes lends some credence to its potential positioning among the transition metals, but, much like lutetium's fellow lanthanoids, it can be used as a dopant in garnet crystals, this time in gadolinium gallium garnet (GGG), which is used in computer memory hardware.

The French connection: Lutetium was named after the Latin name for Paris, the native city of its discoverer, George Urbain.

| 1 | 2 | 3 | 4 | 5 | 6 | 7 | 8 | 9 | 10 | 11 | 12 | 13 | 14 | 15 | 16 | 17 | 18 |
|---|---|---|---|---|---|---|---|---|---|---|---|---|---|---|---|---|---|
| 1 H | | | | | | | | | | | | | | | | | 2 He |
| 3 Li | 4 Be | | | | | | | | | | | 5 B | 6 C | 7 N | 8 O | 9 F | 10 Ne |
| 11 Na | 12 Mg | | | | | | | | | | | 13 Al | 14 Si | 15 P | 16 S | 17 Cl | 18 Ar |
| 19 K | 20 Ca | 21 Sc | 22 Ti | 23 V | 24 Cr | 25 Mn | 26 Fe | 27 Co | 28 Ni | 29 Cu | 30 Zn | 31 Ga | 32 Ge | 33 As | 34 Se | 35 Br | 36 Kr |
| 37 Rb | 38 Sr | 39 Y | 40 Zr | 41 Nb | 42 Mo | 43 Tc | 44 Ru | 45 Rh | 46 Pd | 47 Ag | 48 Cd | 49 In | 50 Sn | 51 Sb | 52 Te | 53 I | 54 Xe |
| 55 Cs | 56 Ba | 71 Lu | 72 Hf | 73 Ta | 74 W | 75 Re | 76 Os | 77 Ir | 78 Pt | 79 Au | 80 Hg | 81 Tl | 82 Pb | 83 Bi | 84 Po | 85 At | 86 Rn |
| 87 Fr | 88 Ra | | 104 Rf | 105 Db | 106 Sg | 107 Bh | 108 Hs | 109 Mt | 110 Ds | 111 Rg | 112 Cn | 113 Nh | 114 Fl | 115 Mc | 116 Lv | 117 Ts | 118 Og |
| | | | 57 La | 58 Ce | 59 Pr | 60 Nd | 61 Pm | 62 Sm | 63 Eu | 64 Gd | 65 Tb | 66 Dy | 67 Ho | 68 Er | 69 Tm | 70 Yb | 71 Lu |
| | | | 89 Ac | 90 Th | 91 Pa | 92 U | 93 Np | 94 Pu | 95 Am | 96 Cm | 97 Bk | 98 Cf | 99 Es | 100 Fm | 101 Md | 102 No | 103 Lr |

There remains a debate over the correct position of lutetium in the periodic table. Should it sit at the end of the lanthanoid series, or should it be a group-3 element?

# ACTINOIDS

**As we shall see, with one or two notable exceptions, the application of the actinoids is almost solely confined to their radioactive nature. Either as instruments of war, or as nuclear fuels, the instability of their nuclei is at the literal center of both their menace and their delight. Harnessing that power has created not only chemical conflict but also immense political conflict. The actinoids are indeed a collection of elements with a significant influence on human history.**


On the following pages:

| | | | |
|---|---|---|---|
| **Ac** | Actinium | **U** | Uranium |
| **Th** | Thorium | **Pu** | Plutonium |
| **Pa** | Protactinium | **Cm** | Curium |


| 1 | 2 | 3 | 4 | 5 | 6 | 7 | 8 | 9 | 10 | 11 | 12 | 13 | 14 | 15 | 16 | 17 | 18 |
|---|---|---|---|---|---|---|---|---|---|---|---|---|---|---|---|---|---|
| 1 H | | | | | | | | | | | | | | | | | 2 He |
| 3 Li | 4 Be | | | | | | | | | | | 5 B | 6 C | 7 N | 8 O | 9 F | 10 Ne |
| 11 Na | 12 Mg | | | | | | | | | | | 13 Al | 14 Si | 15 P | 16 S | 17 Cl | 18 Ar |
| 19 K | 20 Ca | 21 Sc | 22 Ti | 23 V | 24 Cr | 25 Mn | 26 Fe | 27 Co | 28 Ni | 29 Cu | 30 Zn | 31 Ga | 32 Ge | 33 As | 34 Se | 35 Br | 36 Kr |
| 37 Rb | 38 Sr | 39 Y | 40 Zr | 41 Nb | 42 Mo | 43 Tc | 44 Ru | 45 Rh | 46 Pd | 47 Ag | 48 Cd | 49 In | 50 Sn | 51 Sb | 52 Te | 53 I | 54 Xe |
| 55 Cs | 56 Ba | 57-71 | 72 Hf | 73 Ta | 74 W | 75 Re | 76 Os | 77 Ir | 78 Pt | 79 Au | 80 Hg | 81 Tl | 82 Pb | 83 Bi | 84 Po | 85 At | 86 Rn |
| 87 Fr | 88 Ra | **89-103** | 104 Rf | 105 Db | 106 Sg | 107 Bh | 108 Hs | 109 Mt | 110 Ds | 111 Rg | 112 Cn | 113 Nh | 114 Fl | 115 Mc | 116 Lv | 117 Ts | 118 Og |
| | | | 57 La | 58 Ce | 59 Pr | 60 Nd | 61 Pm | 62 Sm | 63 Eu | 64 Gd | 65 Tb | 66 Dy | 67 Ho | 68 Er | 69 Tm | 70 Yb | 71 Lu |
| | | | **89 Ac** | **90 Th** | **91 Pa** | **92 U** | **93 Np** | **94 Pu** | **95 Am** | **96 Cm** | **97 Bk** | **98 Cf** | **99 Es** | **100 Fm** | **101 Md** | **102 No** | **103 Lr** |

## WHAT *EXACTLY* ARE WE DEALING WITH?

The actinoids are a series of fifteen radioactive elements that start with element number 89, actinium, and pass through to element number 103, lawrencium. It is good to establish that fact first, since as with the lanthanoids/lanthanides, throughout history there has been some confusion over the collective name for this band of elements, and even disputes over which elements to include. The first problem arises when one considers the literal meaning of the word *actinoid*. In the strictest sense, actinoid means "like actinium" and therefore should really exclude actinium from the list. So why did we start using a piece of terminology that was problematic in this manner? Well, at least in part because of a *different* problem with the original name for the group, the actinides. When used in chemistry, the *-ide* ending refers to a negative ion such as fluoride F- or chloride Cl-. As such, it was deemed inappropriate for use as a collective term for an alliance of neutral elements. In reality, since both actinoid and actinide have their own issues, and since each has a history of use, both remain acceptable in most contexts, although the official IUPAC Red Book prefers "actinoids."

A closer look at the elements in the set reveals two distinct subsets. First, there are those that occur in nature in relatively significant amounts—this is relative, since none of the actinoids are especially abundant. This list includes thorium, uranium, and protactinium. Then there are the remainder of the actinoids, which are far less abundant and present on earth in only tiny amounts, often only as products of natural radioactive decay or artificially produced in laboratories.

▼ The familiar sight of massive cooling towers at a nuclear power plant. The actinoids are renowned for their radioactive nature and associated applications.

# Actinium

**Chemical symbol**
Ac

**Atomic number**
89

**Atomic mass**
227 (longest-living isotope)

**Boiling point**
5,788°F (3,198°C)

**Melting point**
1,923.8°F (1,051°C)

**Electron configuration**
2.8.18.32.18.9.2

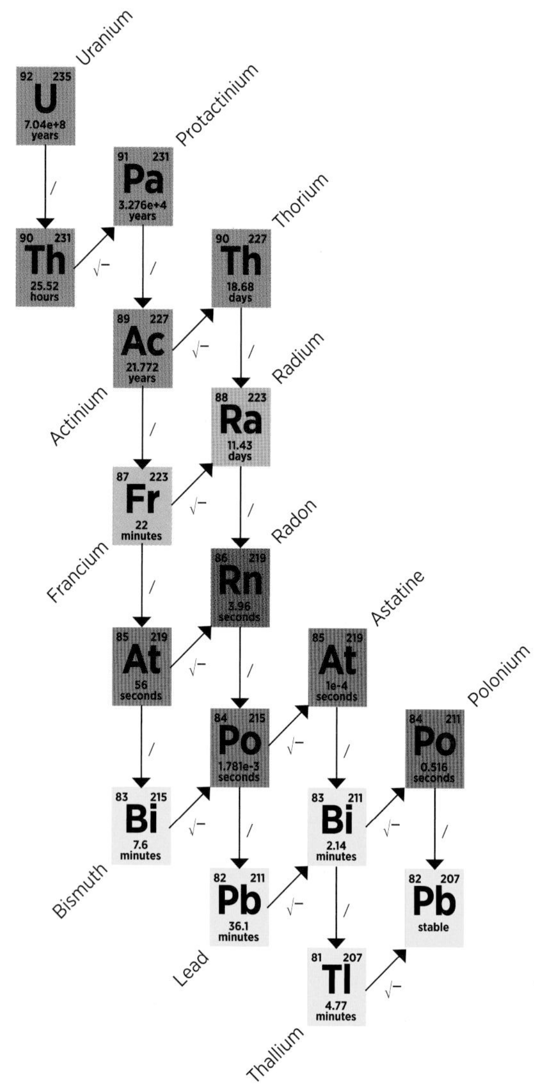

**The current importance of actinium in the world is extremely limited. It is used in a few medical applications. But as the first, and therefore as a representative element in the actinoid series, actinium allows one to make some interesting observations about radioactivity.**

Discovered by André Debierne (1874–1949) in Paris in 1899, any potential that actinium might have had as an element of significant use to mankind was scuppered by two key characteristics. First, it is one of the least abundant elements in the earth's crust. That fact alone means that even if it did not possess the second characteristic, ferocious radioactivity, it would be tricky to put the element to much practical use.

## ALPHA-CELL ATTACK

Actinium has been used in a cancer treatment procedure known as Medical Actinium for Therapeutic Treatment, or MATT. Actinium-225 is an alpha-emitter, as it decays to bismuth-213, emitting alpha particles along the way. The radionuclide can be attached to a targeting agent, which can seek specific cancerous cells, and the alpha particles can then destroy the cancer. The advantages of MATT are that the half-lives of the actinium isotope and its daughter nuclides are short (meaning residual radiation after treatment is not a big concern), and the alpha radiation is not especially penetrating (meaning collateral cell damage is minimal).

- Actinoids
- Alkaline earth metals
- Alkali metals
- Noble gases
- Metalloids
- Post-transition metals

The actinium decay chain shows how nuclei are converted from one element to another by the emission of either alpha or beta particles. Alpha emission reduces the number of protons and hence the atomic number, while beta emission increases each.

## AN UNSTABLE ELEMENT

All of the actinoid elements are radioactive, but in the case of actinium, there are no stable isotopes. This means that every atomic version of the element, no matter how many neutrons are tied up in the nucleus with the consistent eighty-nine protons, is unstable, and will quickly disintegrate into other elements. All of the isotopes of actinium reported have short half-lives (ranging from nanoseconds to a few years), and this contributes to the virtual absence of the element in the earth's crust. It is really only actinium-227, with a half-life of approximately twenty-two years, that we have any significant knowledge of.

As a result of the relatively long half-life of actinium-227, it is the isotope that is most likely to be encountered naturally. What little actinium there is on earth is found in conjunction with the far more abundant actinoid uranium. As a result, any actinium needed for research purposes needs to be prepared artificially by the bombardment of radium-226 with neutrons to give radium-227:

$$^{226}_{88}Ra + ^{1}_{0}n \rightarrow ^{227}_{88}Ra$$

Then the radium-227 decays via beta emission to produce actinium-227:

$$^{227}_{88}Ra \rightarrow ^{227}_{89}Ac + ^{0}_{-1}\surd$$

# Thorium

| Chemical symbol | Th |
|---|---|
| Atomic number | 90 |
| Atomic mass | 232.04 |
| Boiling point | 8,650°F (4,788°C) |
| Melting point | 3,182°F (1,750°C) |
| Electron configuration | 2.8.18.32.18.10.2 |

**Another dangerously radioactive element (and an element that yields a host of dangerously radioactive isotopes through its decay series), thorium now finds use in nuclear reactors, but as the 1930s-marketed health elixir Radithor, it promised (along with radium) some tremendous health benefits.**

Jöns Jakob Berzelius first recognized thorium as an element in 1829, when he identified it as a new substance in a mineral containing thorium silicate. Named after the Norse god of thunder, Thor, thorium had to wait another three-quarters of a century before it came into its own.

### RADITHOR

In the first few years of the twentieth century the recently discovered phenomenon of radioactivity was being touted as a new wonder medicine. Radithor was just one such item in a list of long-since-debunked radioactive quackery.

### BURNING BRIGHT

The element found legitimate use in gas lighting. As one of the elements infused, in its oxide form, into the mantles of gas lamps, the thorium compound would emit a bright, white light.

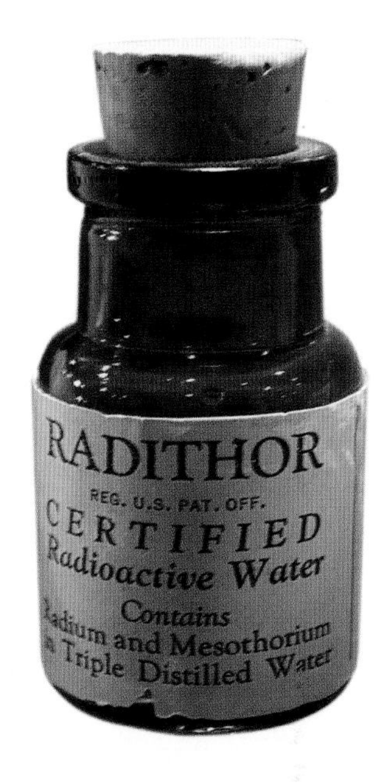


Radithor is an example of a product that falls into the category of radioactive quackery—radioactive materials incorrectly promoted as remedies for various ailments.

# Protactinium

| Chemical symbol | Pa |
|---|---|
| Atomic number | 91 |
| Atomic mass | 231.04 |
| Boiling point | 7,280°F (4,027°C) |
| Melting point | 2,861.6°F (1,572°C) |
| Electron configuration | 2.8.18.32.20.9.2 |

**Eka-tantalum, or if one prefers, maybe brevium, UX2, or perhaps even protoactinium, it doesn't really matter—you are talking about element number 91 whichever name you choose. A complex history of attempted discovery and clumsy naming precedes the modern moniker for element number 91, all fraught with confusion and indecision!**

### WHAT'S IN A NAME?

$UX_2$, aka brevium, was observed by Kazimierz Fajans (1887–1975) and Oswald Helmuth Göhring (1889–c. 1915) in 1913. The name $UX_2$ was the second in a series of names given to particles that originated from the radioactive decay of uranium. The name "brevium" was chosen for $UX_2$ to reflect the short time that the element existed. A few years later, element number 91, or perhaps more accurately, an isotope of element number 91 with a longer half-life, came to the attention of Lise Meitner (1878–1968) and Otto Hahn (1879–1968). In 1918 they published a paper reporting their results that suggested the name "protoactinium" for the "new" element, a name derived from the Greek *protos*, meaning "first." The name didn't exactly roll off the tongue, so the IUPAC settled on the slightly easier to pronounce "protactinium" in 1949.

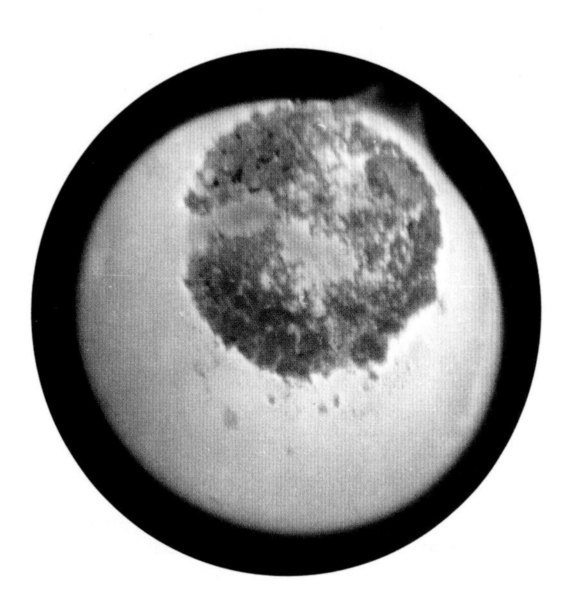

The darker area at the center of the photo is a sample of protactinium-233, photographed in the light from its own radioactive emission (the lighter area), at the National Reactor Testing Station in Idaho, circa 1969.

# Uranium

| Chemical symbol | U |
|---|---|
| Atomic number | 92 |
| Atomic mass | 283.03 |
| Boiling point | 7,468°F (4,131°C) |
| Melting point | 2,070°F (1,132°C) |
| Electron configuration | 2.8.18.32.21.9.2 |

Henri Becquerel's contribution to our understanding of radioactivity is immortalized in the name of the SI unit for radioactivity: the becquerel (Bq).

**After a pretty slow start to its life as an element, uranium became a vitally important one, since it helped to birth the Nuclear Age. Uranium's influence truly began in 1896, with the discovery of radioactivity. Following this, uranium went from being an element with little significance in the grand scheme, to arguably one of the most important of the modern age.**

After Martin Heinrich Klaproth discovered uranium in 1789, not much happened with the element until, in 1896, the French chemist Henri Becquerel (1852–1908) discovered radioactivity as a result of the spontaneous decay of some uranium salts. Becquerel was investigating phosphorescence and X-rays when he noticed that photographic plates that had been exposed to the uranium salts but no sunlight still developed images. He concluded that the uranium salts were emitting some form of radiation. In the 1930s and '40s, work by various scientists, both inside and outside the U.S. government's Manhattan Project (the research project designed to develop nuclear weapons), helped to cement uranium's place as an element of vital importance, when the potential for nuclear fission was proposed.

## LITTLE BOY

For use in the Little Boy nuclear bomb that was dropped on Hiroshima, Japan, on the morning of August 6, 1945, American scientists needed one particular isotope of uranium, U-235. U-235 is the isotope that, when bombarded with neutrons, splits apart to form lighter nuclei and more neutrons. This process, known as fission, produces energy, and since neutrons are produced in the process, it can initiate a chain reaction that releases almost unfathomable amounts of energy. The uncontrolled release of that energy is the essence of a nuclear bomb. Uranium is relatively plentiful on earth, being the forty-sixth most abundant element, and exists as three naturally occurring isotopes: U-238, U-235, and U-234. However, the U-238 isotope makes up over 99 percent of the supply, and therefore, before a critical mass (the mass required to sustain the chain reaction) of the 235 isotope can be obtained, it must go through a process of enrichment to increase the percentage of U-235 present in any given sample.

## FIESTAWARE

Long before its use in nuclear weapons and as a nuclear fuel, compounds of uranium were originally added to pottery and glass in order to produce colorful glazes. In fact, a popular and highly collectible dinnerware line called Fiesta was manufactured in the United States using a number of such glazes. The orange-red colors of Fiesta dinnerware used to contain significant amounts of uranium oxides, and this led to the cups, saucers, and plates having detectable radioactivity counts!

▲ A model of Little Boy, the atomic bomb that was dropped on Hiroshima from the B-29 Superfortress aircraft *Enola Gay*.

◀ Some glazes used for the twentieth-century dinnerware known as Fiestaware were manufactured using uranium oxide and have been found to be radioactive.

# Plutonium

| Chemical symbol | Pu |
|---|---|
| Atomic number | 94 |
| Atomic mass | 244 (longest-living isotope) |
| Boiling point | 5,842°F (3,228°C) |
| Melting point | 1,183°F (639.5°C) |
| Electron configuration | 2.8.18.32.24.8.2 |

**Like uranium, plutonium is without doubt an element that impacted world history. As a fissile material, Pu-239 can provide a sustainable nuclear chain reaction, releasing enormous amounts of energy in the process. The Fat Man plutonium bomb dropped on the Japanese city of Nagasaki on August 9, 1945, three days after the uranium-based Little Boy, effectively ended World War II.**

A model of Fat Man, the atomic bomb that was dropped on Nagasaki from the B-29 Superfortress aircraft *Bockscar*.

## PLANETARY BODY

The element was first synthesized in Berkeley, California, in late 1940. However, due to national security concerns, the Americans wanted to prevent word of its discovery from getting out, and by the time it was formally announced in 1946, Nagasaki had already been flattened. Plutonium is the third element in a series that were named after planets (uranium after Uranus and neptunium after Neptune both precede Pu). There is virtually no plutonium present in the earth's crust, so it has to be manufactured in nuclear facilities. In the early 1940s, the U.S. government needed a significant amount of plutonium in order to build a nuclear bomb, so they set up a production facility to make element number 94. The nuclear reactions that they used are shown below.

$$^{238}_{92}U + ^{1}_{0}n \rightarrow ^{239}_{92}U$$

$$^{239}_{92}U \rightarrow ^{239}_{93}Np + ^{0}_{-1}\surd$$

$$^{239}_{93}Np \rightarrow ^{239}_{94}Pu + ^{0}_{-1}\surd$$

In the first reaction, U-238 captures a neutron to become the heavier isotope, U-239. In the second reaction, that isotope emits a beta particle to become an isotope of neptunium, element 93. In the final nuclear event, another beta particle is released, and neptunium is converted to Pu-239. It's this isotope of plutonium that was used in the Fat Man nuclear bomb.

## PEACETIME PLUTONIUM

In times of relative peace, the nuclear power of plutonium is harnessed for the production of nuclear energy. One type of nuclear reactor uses a fuel that is a compound of plutonium, plutonium dioxide ($PuO_2$). Combined with the equivalent oxide of uranium, $UO_2$, the fuel is commonly known as MOX, or mixed oxide fuel.

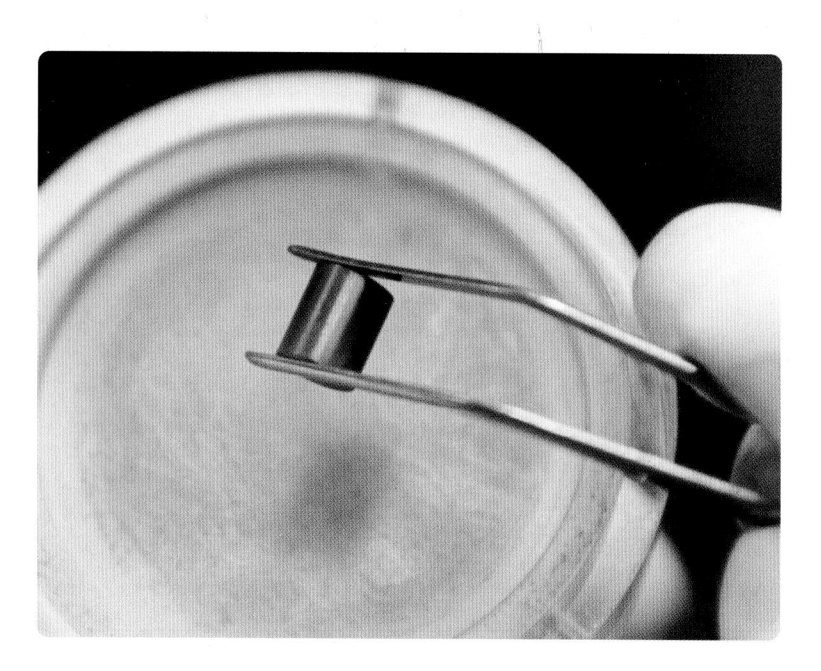

Nuclear fuel pellets like this one are often made of MOX, mixed oxides of plutonium and uranium. The uranium and plutonium in the pellets are the source of the nuclear reaction that takes place in the reactor.

# Curium

| | |
|---|---|
| **Chemical symbol** | Cm |
| **Atomic number** | 96 |
| **Atomic mass** | 247 (longest living isotope) |
| **Boiling point** | 5,612°F (3,100°C) |
| **Melting point** | 2,453°F (1,345°C) |
| **Electron configuration** | 2.8.18.32.25.9.2 |

## KEY FIGURE

**MARIE CURIE** 1867–1934

Native Pole Marie Curie and her French husband Pierre were pioneering scientists in the field of radioactivity. They were awarded the 1903 Nobel Prize in Physics for their work. In 1911 Marie Curie was awarded the Nobel Prize in Chemistry for her discoveries of radium and polonium. In the process, she became the first person to receive two Nobel Prizes.

**Curium is noteworthy if for no other reason than that it is named (in part) after the woman who discovered two other elements, Marie Curie. I say "in part," since formally element number 96 is dedicated to both her and her husband, Pierre, but Marie's impact as a female in the otherwise almost exclusively male-dominated world of the history of radioactivity, the discovery of the elements, and the periodic table in general, is worthy of particular recognition.**

The ultimate prize for any element-seeker is to have an element named after himself or herself. In the early days of elemental discovery there was a tremendous amount of skulduggery and downright lying and cheating. In 1947, the IUPAC stepped in and said enough is enough. The rules that they initially established for naming have evolved over time, to the point where new elements must now be named after one of the following:

- a mythological concept or character (including an astronomical object)
- a mineral or similar substance
- a place or geographical region
- a property of the element
- a scientist

## A PREMATURE ANNOUNCEMENT

When, in 1944, curium was synthesized for the first time, no such specific rules existed, but that was the least of the problems associated with curium's announcement to the world. Curium was found as a result of research into nuclear weapons in the context of the Manhattan Project and World War II. As such, discoveries were bound by concerns over American national security, and announcements of new elements via the usual channels of scientific journals had to wait. In fact, in the end, the co-discoverer of curium, Glenn Seaborg, made a casual announcement on a children's radio program when a kid asked him if any new elements had recently been discovered. He answered in a somewhat matter-of-fact manner by saying, "Why, yes, yes they have!"

## AN ELEMENT OF FEW USES

Outside of its historical significance, curium's practical applications are extremely limited. As an alpha particle emitter, it has a few uses (curium-242 can be used as a source of thermal energy since its decay is associated with an unusually high production of heat), but the unpleasant radioactive nature of most of its decay products means that it is not suitable for any extensive applications.

▲ Glenn Seaborg had an extraordinary career in nuclear chemistry, and in particular in the synthesis of new elements. He was honored by having element number 106, seaborgium, named after him.

# NONMETALS

Much like the term "metalloid," the definition of a nonmetal is somewhat ambiguous. The elements typically distinguished as nonmetals exhibit a very wide range of properties. They include solids, liquids, and gases; they encompass elements from multiple groups on the periodic table; and they show a wide range of chemical properties. Given this diversity, perhaps it is understandable that we tend to use defining criteria for nonmetals that tell us what these elements *are not*, rather than what they *are*.

On the following pages:

| | | | |
|---|---|---|---|
| **C** | Carbon | **P** | Phosphorus |
| **N** | Nitrogen | **S** | Sulfur |
| **O** | Oxygen | **Se** | Selenium |

| 1 | 2 | 3 | 4 | 5 | 6 | 7 | 8 | 9 | 10 | 11 | 12 | 13 | 14 | 15 | 16 | 17 | 18 |
|---|---|---|---|---|---|---|---|---|---|---|---|---|---|---|---|---|---|
| 1 H | | | | | | | | | | | | | | | | | 2 He |
| 3 Li | 4 Be | | | | | | | | | | | 5 B | **6 C** | **7 N** | **8 O** | 9 F | 10 Ne |
| 11 Na | 12 Mg | | | | | | | | | | | 13 Al | 14 Si | **15 P** | **16 S** | 17 Cl | 18 Ar |
| 19 K | 20 Ca | 21 Sc | 22 Ti | 23 V | 24 Cr | 25 Mn | 26 Fe | 27 Co | 28 Ni | 29 Cu | 30 Zn | 31 Ga | 32 Ge | 33 As | **34 Se** | 35 Br | 36 Kr |
| 37 Rb | 38 Sr | 39 Y | 40 Zr | 41 Nb | 42 Mo | 43 Tc | 44 Ru | 45 Rh | 46 Pd | 47 Ag | 48 Cd | 49 In | 50 Sn | 51 Sb | 52 Te | 53 I | 54 Xe |
| 55 Cs | 56 Ba | 57-71 | 72 Hf | 73 Ta | 74 W | 75 Re | 76 Os | 77 Ir | 78 Pt | 79 Au | 80 Hg | 81 Tl | 82 Pb | 83 Bi | 84 Po | 85 At | 86 Rn |
| 87 Fr | 88 Ra | 89-103 | 104 Rf | 105 Db | 106 Sg | 107 Bh | 108 Hs | 109 Mt | 110 Ds | 111 Rg | 112 Cn | 113 Nh | 114 Fl | 115 Mc | 116 Lv | 117 Ts | 118 Og |
| | | | 57 La | 58 Ce | 59 Pr | 60 Nd | 61 Pm | 62 Sm | 63 Eu | 64 Gd | 65 Tb | 66 Dy | 67 Ho | 68 Er | 69 Tm | 70 Yb | 71 Lu |
| | | | 89 Ac | 90 Th | 91 Pa | 92 U | 93 Np | 94 Pu | 95 Am | 96 Cm | 97 Bk | 98 Cf | 99 Es | 100 Fm | 101 Md | 102 No | 103 Lr |

## METAL OPPOSITES

We typically think of the nonmetals as poor conductors of heat and electricity, as elements that tend to either gain or share electrons during chemical reactions, that have relatively low melting and boiling points, and that are not malleable or ductile. By using these parameters, what we are really saying is that they possess many properties that are the opposite of metals. In this regard the classification makes sense, but there are problems with such flexibility. For example, although these general ideas commonly create a list of seventeen elements that we classify as nonmetals (hydrogen, helium, nitrogen, oxygen, fluorine, neon, chlorine, argon, krypton, xenon, radon, bromine, carbon, phosphorus, sulfur, selenium, and iodine), there is wiggle room. Selenium may be classified as a metalloid, and further subtle nuances also occur. In this book, as in many similar instances, several of the nonmetals in that list of seventeen are extracted into their own collections (the halogens and the noble gases, for example) for separate discussion. It makes sense for them to be treated in that way because their similarities with members of their respective groups would be lost and not highlighted if they were collected together in the much more general category of nonmetals. Additionally, pulling hydrogen out as a single entity, unassociated with any other elements, best highlights its uniqueness.

Those considerations leave us with a list of six elements that we will treat as the quintessential nonmetals: carbon, nitrogen, oxygen, phosphorus, sulfur, and selenium.

**A mixture of gases, chiefly sulfur dioxide, billowing into the sky from a sulfur mine in East Java, Indonesia.**

# Carbon

| | |
|---|---|
| **Chemical symbol** | C |
| **Atomic number** | 6 |
| **Atomic mass** | 12.011 |
| **Boiling point** | (graphite) 6,422°F (3,550°C) |
| **Melting point** | (graphite) 6,917°F (3,825°C) |
| **Electron configuration** | 2.4 |

**Is element number 6 the most important element on earth? One can construct a pretty convincing argument in favor of answering yes. As the basis of all of organic chemistry, and hence all chemistry of life, carbon is literally at the center of the most important molecules known. Whether it is in DNA and chlorophyll, or amino acids and cellulose, plant and animal life is constructed around, and dependent on, molecules made from carbon atoms.**

## EARLY DAYS

Carbon can be considered as being one of the prehistoric elements, inasmuch as no single discoverer can be easily defined. As humans were playing around with fire, and inevitably charcoal, they were handling pure carbon. Lavoisier's original list of elements, published in 1789, included carbon, as did Dalton's list published in 1805.

▲ Organic chemistry is the chemistry of carbon. Carbon atoms form the backbone of a multitude of naturally occurring, vitally important compounds, such as cellulose and chlorophyll, found in plants.

▶ Charcoal is a familiar manifestation of carbon that we may encounter in our everyday lives. It can be produced from the slow heating of wood in the absence of oxygen.

## ISOTOPIC SIGNIFICANCE

As the fifteenth most abundant element on earth, carbon exists naturally as three isotopes, carbon 12, carbon 13, and carbon 14. Two of these have special significance in chemistry, and for very different reasons. Carbon-12 is used as the standard by which the masses of all other atoms are measured. Taking the mass of 1/12 of a carbon-12 atom as a standard, all atoms have masses expressed as multiples of that unit. Carbon-14 is radioactive, with a half-life of 5,730 years. By measuring the amount of carbon-14 remaining in dead plant or animal material, the age of that material can be calculated. The older the material, the less carbon-14 will remain.

## FULLERENES

With its unique ability to form four bonds with itself, and its two familiar allotropes of graphite and diamond, carbon truly is an extraordinary element. In the mid-1980s and into the 1990s, a further allotrope of carbon became known. In addition to the much more familiar diamond and graphite, a number of other, more elaborate, hollow three-dimensional arrangements of carbon atoms called fullerenes were prepared, first at Rice University in Texas. Having been proposed by a number of scientists around the world prior to the 1980s, the first one produced was a soccer ball–shaped structure composed of carbon atoms arranged in hexagonal and pentagonal formations. It is known as buckminsterfullerene, $C_{60}$.

Crystallized buckminsterfullerene, $C_{60}$, one of the allotropic forms of elemental carbon.

Radiocarbon dating is used extensively in archaeology to date artifacts containing organic material. It involves measuring the amount of carbon-14 present in the objects and using the isotope's half-life to determine their age.

# Nitrogen

| Chemical symbol | N |
|---|---|
| Atomic number | 7 |
| Atomic mass | 14.007 |
| Boiling point | -320.43°F (-195.8°C) |
| Melting point | -346°F (-210°C) |
| Electron configuration | 2.5 |

**Any element that makes up approximately 78 percent of the Earth's atmosphere must be pretty important, and element number 7 is certainly a significant one. Like oxygen, nitrogen forms a diatomic molecule in its elemental state. The $N_2$ molecule differs from the $O_2$ molecule in one significant way, though. While oxygen atoms are bonded to one another with a double covalent bond, the two nitrogen atoms in an $N_2$ molecule are bound together by a triple bond.**

The oxygen molecule is formed when a pair of oxygen atoms are covalently bonded with a double bond. The nitrogen atoms in $N_2$, on the other hand, make a triple covalent bond.

Scottish chemist Daniel Rutherford was a student of Joseph Black. In 1772, Rutherford was the first person to isolate nitrogen. He called it "noxious air."

## NOXIOUS AIR

With so much of it floating around in the atmosphere, and an enormous amount of investigation of "airs" (the collective name for various gases in the second half of the eighteenth century), nitrogen's relatively early discovery was inevitable. Several chemists came close, but the discovery is credited to Daniel Rutherford (1749–1819), a Scottish chemist. His experimentation essentially involved suffocating a mouse. The mouse was confined to a small container where it used up all of the oxygen. Rutherford realized that there was still "an air" left in the container that would not support respiration, and since it caused the death of the mouse, he called it "noxious" air.

## EXPLOSIVE BOND

As a gas, diatomic nitrogen is basically inert. This is due mainly to the incredibly strong triple bond that two nitrogen atoms form in an $N_2$ molecule. Many compounds contain nitrogen atoms that are not combined with other nitrogen atoms, for example in trinitrotoluene, better known as the explosive TNT. TNT can be easily provoked into a reaction where the individual nitrogen atoms in the compounds are persuaded to come together and form $N_2$ gas. The incredible force by which the atoms are brought together to make the diatomic molecule releases a large amount of energy. When this process is repeated in a typical reaction trillions of times, the release of energy is colossal, and this is how nitrogen-based explosives work. This power is also harnessed in automobile airbags, which contain sodium azide, $NaN_3$. When detonated in a collision, $NaN_3$ also forms large amounts of $N_2$ extremely rapidly, causing the airbag to inflate.

## NITROGEN FIXATION

Nitrogen has a crucial role to play in biology in such essential compounds as amino acids, and plants need to be able to "fix" nitrogen by taking relatively useless $N_2$ and converting it to useful compounds such as ammonia, $NH_3$. Nitrogen fixation can occur naturally when bacteria in the soil convert nitrogen for plant use, but many fertilizers such as ammonium nitrate ($NH_4NO_3$) are important agricultural chemicals, since they provide nitrogen in a usable form.

Arguably the most important application of nitrogen is in the manufacture of fertilizers and, hence, the production of food for the planet.

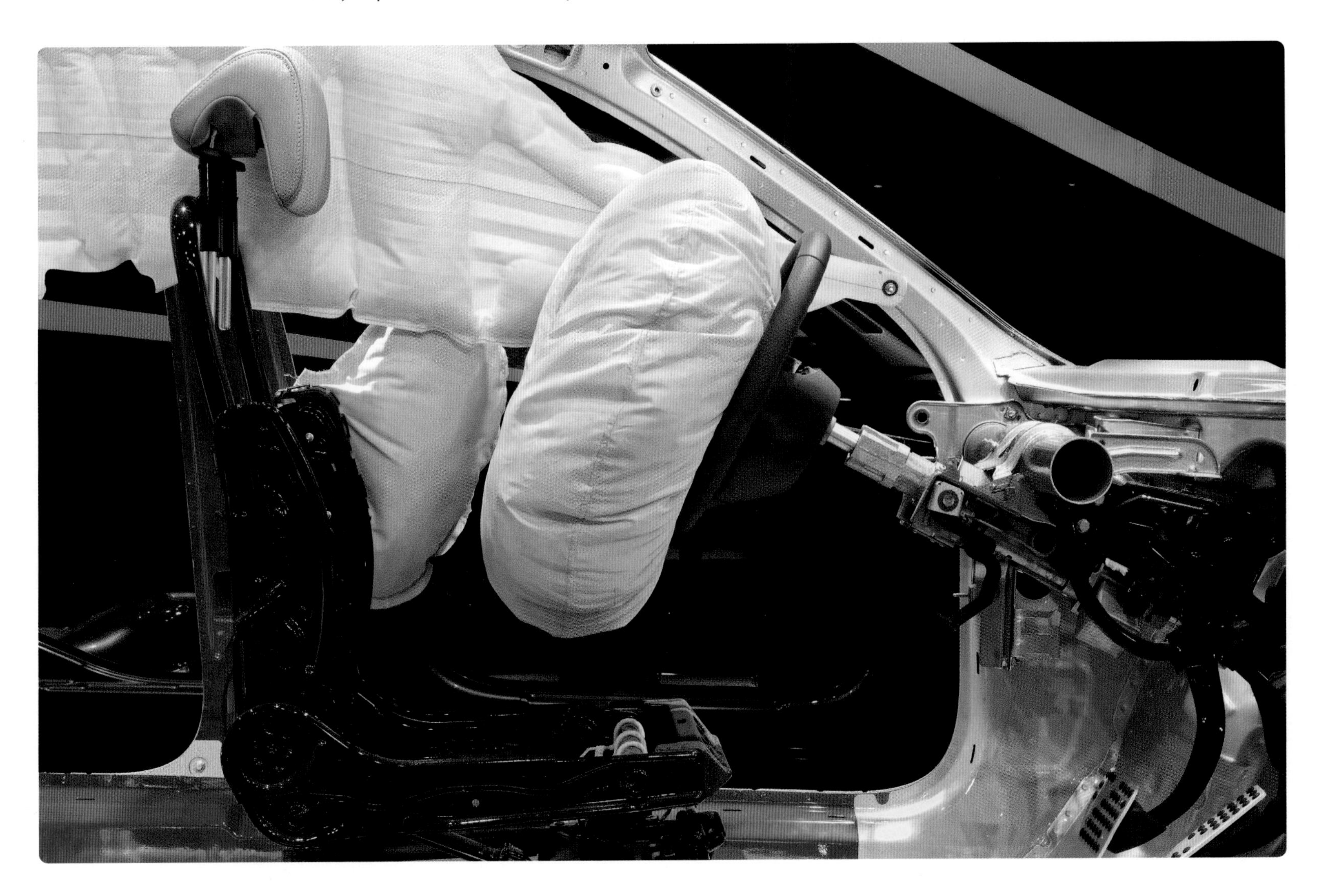

It is the instantaneous release of a controlled amount of nitrogen gas in a chemical reaction that allows an airbag to inflate rapidly.

# Oxygen

| | |
|---|---|
| **Chemical symbol** | O |
| **Atomic number** | 8 |
| **Atomic mass** | 15.999 |
| **Boiling point** | -297.33°F (-182.96°C) |
| **Melting point** | -361.82°F (-218.79°C) |
| **Electron configuration** | 2.6 |

"The feeling of it [oxygen] to my lungs was not sensibly different from that of common air, but I fancied that my breast felt peculiarly light and easy for some time afterwards."

—Joseph Priestley, *Experiments and Observations on Different Kinds of Air*, 1775

**If its fellow nonmetal element carbon is the very stuff of life, then the gaseous element just a couple of spots to the right on the periodic table isn't far behind in terms of its importance—it drives life. Without it, animals, including humans, could not survive, so its ubiquitous presence in the atmosphere is the very key to life on earth.**

### WHO GETS THE CREDIT?

In addition to making up 21 percent of the atmosphere of Earth, oxygen is the third most abundant element in the universe as a whole. As such, it was destined to be one of the elements that would be discovered early in the history of chemistry. The discovery (and discoverer) of oxygen is open to debate. The Swede Carl Wilhelm Scheele first identified the gas in 1772 by liberating it from a number of oxygen-containing compounds, but his findings were not published until a few years later. In the meantime, Englishman Joseph Priestley (1733–1804) identified oxygen via the decomposition of an oxide of mercury, and since he published his findings first, he is often awarded priority. Priestley experimented with the new gas by exposing both mice and himself to it. Depending on the source, Priestley, Scheele, or even both, are sometimes credited with the discovery of the first member of group 16.

### IN NATURE AND IN INDUSTRY

Because of its central role to respiration, it is easy to see why oxygen is so important, but the importance of element number 8 to life on Earth goes way beyond that. As one of its allotropes, ozone ($O_3$), oxygen protects the Earth from the Sun's harmful ultraviolet rays. The depletion of the ozone layer around the Earth by reaction with many chemicals, notably CFCs, has led to the banning of many such compounds. As a relatively soluble gas, oxygen is found in water systems, and helps to sustain a large amount of aquatic animal and plant life. Oxygen is a crucial substance in many industrial processes, not least of all as it supports combustion (burning). Whenever a situation requires the burning of a fuel, oxygen must be present, and in industrial settings its addition can help increase the temperatures at which things burn.

### OXIDATION

Despite its vital role in life, oxygen's ubiquitous presence in the atmosphere is not always welcome. As a reactive gas, over time oxygen tends to react with just about anything that it comes into contact with. This is especially true of its reaction with metals, notably iron. Such a reaction, known generically as oxidation, may simply produce a relatively harmless oxide layer that tarnishes the surface of the metal—as it does when silver cutlery is exposed to air—or it may pose a more serious problem, leading to mechanical corrosion when iron rusts.

Approximately 21 percent of the air in Earth's atmosphere is oxygen, the remainder being almost exclusively nitrogen, with small amounts of several other gases also present.

Rusting iron nails exhibit the characteristic reddish-brown hydrated iron(III) oxide coating that we associate with the corroding metal.

# Phosphorus

| | |
|---|---|
| **Chemical symbol** | P |
| **Atomic number** | 15 |
| **Atomic mass** | 30.974 |
| **Boiling point** | (white) 536.9°F (280.5°C) |
| **Melting point** | (white) 111.47°F (44.15°C) |
| **Electron configuration** | 2.8.5 |

**Phosphorus's reputation as an element of chaos often precedes it, so it can be easy to overlook the crucial role that it plays in living cells. As an element of war, element number 15 has been responsible for a great deal of death and destruction. As an element of life, its role in the vital biological molecules ATP (adenosine triphosphate) and DNA means that we simply cannot survive without it.**

### THE PHILOSOPHER'S STONE

Hennig Brand (1630–1710) was one of many alchemists of his time who were searching for the philosopher's stone. A legendary substance, the philosopher's stone was thought to possess a number of extraordinary powers that included the ability to deliver immortality and to turn base metals such as mercury and lead into gold. The idea of the philosopher's stone spurred all kinds of astonishing, chemical wild-goose chases, and it was on one of these that Brand discovered phosphorus in 1669. By heating a vat of human urine, Brand had hoped ultimately to extract gold. What he got instead was a substance that glowed in air, and that he called *phosphorus* after the same Greek word, meaning "bearer of light."

### DEEP BURN

Phosphorus is another of the nonmetallic elements that can exist in a number of allotropic forms. The different forms are characterized by their colors, with white, red, and black being three common types. It is the flammable and poisonous white form that has had the most devastating effects in times of war, as it has been used to create smoke screens and terrifying incendiary bombs. White phosphorus is known to cause deep second- and third-degree burns on victims, and is a devastating poison—the aftereffects of a bombardment with white phosphorus include severe liver problems.

### THE DEVIL'S ELEMENT

The incendiary nature of white phosphorus was harnessed by its use in the match industry in the late nineteenth century. The workers in match factories were subject to dreadful working conditions, including the horrible occupational disease "phossy jaw." Caused once again by the nefarious white phosphorus used to make the matches, the condition created painful and disfiguring abscesses in the jawbone and a putrid-smelling pus. Without treatment, the disease could prove fatal.

◀ It has been suggested that Joseph Wright's famous painting of 1771, *The Alchymist in Search of the Philosopher's Stone Discovers Phosphorus*, refers to Hennig Brand's discovery of phosphorus in 1669.

The United States used white phosphorus munitions against the Vietcong during the Vietnam War.

The London match girls' strike of 1888 was an early piece of industrial action that protested the dangerous conditions encountered in match factories that used phosphorus.

# Sulfur

**Chemical symbol**
S

**Atomic number**
16

**Atomic mass**
32.06

**Boiling point**
832.3°F (444.6°C)

**Melting point**
239.38°F (115.21°C)

**Electron configuration**
2.8.6

**Mentioned in the Bible multiple times under its pseudonym brimstone, sulfur was also known to the ancient Greeks. In modern chemical nomenclature the prefix *thio-* is used to denote the presence of the element, and "thio" also has a theological connection, being from the Greek *theos*, meaning "god."**

One of the earliest references to sulfur was in the alchemists' belief that it was part of the philosopher's stone. The philosopher's stone was, in its simplest manifestation, a mythical substance that, it was believed, could be used to turn base metals into gold.

In the twenty-first century, sulfur is far more likely to render itself in important compounds used in industry and beyond. Examples include sulfuric acid, one of the world's most important industrial chemicals; sulfur dioxide, which can be used as a preservative and bleaching agent; and in a far more pungent way, methanethiol, aka methyl mercaptan, which is used in various industrial settings including in the manufacture of pesticides. This chemical has the dubious distinction of being one of the foulest smelling on the planet.

## ALLOTROPES

As a member of group 16, the bright yellow solid sits directly underneath oxygen on the periodic table and above the far less ubiquitous selenium. It is a reactive nonmetal with a dizzying array of isotopes and, as a component of some amino acids, is one of the elements that are essential for life.

One of the more recognizable elements, sulfur is perhaps best known via one of its physical properties: that of the powdery, bright yellow substance that many encounter in the high-school chemistry lab. However, this is only one of many different forms (or allotropes) of the element.

## THE WRITE STUFF

If you are a user of British English, the name of this element is likely to raise some hackles. Even if the British spelling of "sulphur" is the one that you are most familiar with, the organization charged with sorting out such disputes, the IUPAC, has long since made the "f" spelling the official one. In 1992, even the quintessentially British Royal Society of Chemistry (RSC) gave up on the "ph" spelling.

Orthorhombic sulfur is the most common crystalline allotrope of sulfur because it is the most stable form below 194°F (96°C).

One of the most visually distinctive elements, large mounds of sulfur can often be seen close to the point of mining.

# Selenium

**Chemical symbol**
Se

**Atomic number**
34

**Atomic mass**
78.971

**Boiling point**
1,265°F (685°C)

**Melting point**
428.9°F (220.5°C)

**Electron configuration**
2.8.18.6

**A somewhat obscure element, selenium is surprisingly one of the essential elements for humans. However, it is a delicate balance: With too little selenium, serious health issues can result, as cells find themselves unable to defend against disease; with too much selenium, there is the risk of a toxic reaction. Similar to the effect of tellurium, bad breath and body odor are symptoms of excess selenium.**

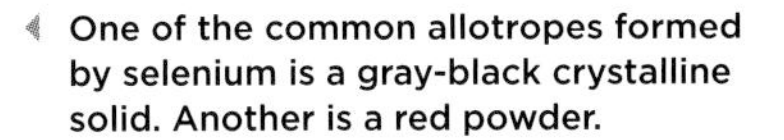

One of the common allotropes formed by selenium is a gray-black crystalline solid. Another is a red powder.

When Jöns Jacob Berzelius discovered selenium in 1817, he named it after the Greek *selene*, meaning "moon." A strange choice, perhaps, but when one knows that tellurium had previously been named after the Latin *tellus*, meaning "earth," his designation makes more sense, as it forever links the two chemically similar elements.

The delicate balance that selenium must find for good health in its role as an antioxidant can be illustrated by a couple of diverse examples: one where deficiencies caused deadly problems, and one where an excess of the element caused a curious observation among grazing animals.

## TOO LITTLE

In the 1960s, it was noticed that disproportionately large numbers of the population in certain regions of China were afflicted with debilitating heart conditions. Most badly affected were children and young women, who were dying from heart failure with worrying regularity. The common thread that ran through the areas where the higher occurrences were found was the selenium-deficient soil.

## TOO MUCH

Selenium can have a surprising effect on livestock. Once again, the chemical makeup of the soil is the key, but in these instances, rather than there being a lack of selenium in the soil, the exact opposite is true. Certain plants can take up large quantities of selenium, so that it becomes concentrated within them. The selenium-rich plants are then eaten by livestock, and the result, known in some places as the blind staggers, can be quite a sight. Cattle have been known to act as if they were drunk, unable to keep their balance and prone to falling over. One plant known to cause the problem, vetch, is also known as locoweed, from the Spanish *loco*, meaning "crazy."

Selenium can accumulate in locoweed, causing problems for livestock that consume the plant.

# HALOGENS

**The halogens consist of the six elements in group 17 of the periodic table: fluorine, chlorine, bromine, iodine, astatine, and the newly minted (in 2016) tennessine. As volatile elements that are seeking one more electron in order to complete their valence p-subshells, they tend to react easily and form many compounds, and their collective name is derived from their willingness to do just that with positive metal ions: it means "salt-forming."**

On the following pages:

| | | | |
|---|---|---|---|
| **F** | Fluorine | **I** | Iodine |
| **Cl** | Chlorine | **At** | Astatine |
| **Br** | Bromine | | |

| 1 | 2 | 3 | 4 | 5 | 6 | 7 | 8 | 9 | 10 | 11 | 12 | 13 | 14 | 15 | 16 | 17 | 18 |
|---|---|---|---|---|---|---|---|---|---|---|---|---|---|---|---|---|---|
| 1 H | | | | | | | | | | | | | | | | | 2 He |
| 3 Li | 4 Be | | | | | | | | | | | 5 B | 6 C | 7 N | 8 O | **9 F** | 10 Ne |
| 11 Na | 12 Mg | | | | | | | | | | | 13 Al | 14 Si | 15 P | 16 S | **17 Cl** | 18 Ar |
| 19 K | 20 Ca | 21 Sc | 22 Ti | 23 V | 24 Cr | 25 Mn | 26 Fe | 27 Co | 28 Ni | 29 Cu | 30 Zn | 31 Ga | 32 Ge | 33 As | 34 Se | **35 Br** | 36 Kr |
| 37 Rb | 38 Sr | 39 Y | 40 Zr | 41 Nb | 42 Mo | 43 Tc | 44 Ru | 45 Rh | 46 Pd | 47 Ag | 48 Cd | 49 In | 50 Sn | 51 Sb | 52 Te | **53 I** | 54 Xe |
| 55 Cs | 56 Ba | 57-71 | 72 Hf | 73 Ta | 74 W | 75 Re | 76 Os | 77 Ir | 78 Pt | 79 Au | 80 Hg | 81 Tl | 82 Pb | 83 Bi | 84 Po | **85 At** | 86 Rn |
| 87 Fr | 88 Ra | 89-103 | 104 Rf | 105 Db | 106 Sg | 107 Bh | 108 Hs | 109 Mt | 110 Ds | 111 Rg | 112 Cn | 113 Nh | 114 Fl | 115 Mc | 116 Lv | **117 Ts** | 118 Og |
| | | | 57 La | 58 Ce | 59 Pr | 60 Nd | 61 Pm | 62 Sm | 63 Eu | 64 Gd | 65 Tb | 66 Dy | 67 Ho | 68 Er | 69 Tm | 70 Yb | 71 Lu |
| | | | 89 Ac | 90 Th | 91 Pa | 92 U | 93 Np | 94 Pu | 95 Am | 96 Cm | 97 Bk | 98 Cf | 99 Es | 100 Fm | 101 Md | 102 No | 103 Lr |

## SHORT-LIVED

Of these six, the first four halogens are by far the most important. The final two are little more than theoretical curiosities due to their radioactive nature, and the fact that no more than a handful of atoms of either have ever been produced. The atoms of astatine and tennessine that have been produced in laboratories are fleeting to say the least, with even the longest-lived isotope of astatine only enjoying a half-life of just over eight hours.

## SALT-PRODUCING

It took just over 100 years, from 1774 to 1886, to discover the first four halogens. Chlorine came first, followed by iodine, bromine, and eventually fluorine. The designation "salt-producing" came from the observation that the elements readily form salts with metals. The most common and ubiquitous of those salts is sodium chloride, but hundreds of such compounds can be made by combining metals with fluorine, chlorine, bromine, and iodine. The collective name for these anions is "halides," or more specifically, fluoride, chloride, bromide, and iodide.

## THE QUEST FOR ELECTRONS

As elements, the halogen atoms pair up with one another and exist as the diatomic molecules $F_2$, $C_{12}$, $Br_2$, and $I_2$. In this form, the elements exhibit increasingly high melting and boiling points, and as a result, their states at room temperature range from gases (fluorine and chlorine), through to a liquid (bromine), and finally to a solid (iodine). All are highly reactive, with fluorine and chlorine particularly so.

Much of the chemistry of the halogens revolves around their desire to gain an electron in order to achieve the more stable electron structure of a filled valence p-subshell. With an outer electronic configuration that sees five of the six spots in the p-subshell filled, the constant quest for electrons to fill the subshell and achieve relative stability makes the halogens excellent oxidizing agents.

**Fluorine, chlorine, bromine, and iodine samples at room temperature. $F_2$ and $Cl_2$ exist as gases, whereas $Br_2$ is a liquid that vaporizes easily, and $I_2$ is a solid that sublimes.**

# Fluorine

| | |
|---|---|
| **Chemical symbol** | F |
| **Atomic number** | 9 |
| **Atomic mass** | 18.998 |
| **Boiling point** | -306.62°F (-188.11°C) |
| **Melting point** | -363.41°F (-219.67°C) |
| **Electron configuration** | 2.7 |

**Sitting at the top of group 17, fluorine is another element with a nasty reputation for being cruel and unforgiving. Frankly, that reputation is well deserved. Throughout the history of the search for fluorine there were many casualties. Both as the highly reactive element and as the stunningly dangerous hydrofluoric acid (HF), element number 9 managed to poison, and ultimately kill, several prominent scientists of the nineteenth century.**

Fluorine was at last tamed in 1886, when Frenchman Henri Moissan finally isolated it. It had been a long journey, and fluorine had wreaked plenty of destruction along the way, including inflicting illness on the great Humphry Davy and the Knox brothers from the Royal Irish Academy, and even killing Paulin Louyet (1818–50) and Jérôme Nicklès (1820–69). Bearing that in mind, Moissan's award of the 1906 Nobel Prize in Chemistry, in part for "his investigation and isolation of the element fluorine," was well deserved!

Long before fluorine was recognized as an element in its own right, it was known to be part of the important mineral fluorspar, $CaF_2$. Fluorspar had been used for centuries in the production of metals, since its addition helped the metals to become fluid. Knowing that, it is not surprising to learn that the name "fluorine" comes from the Latin *fluere*, meaning "to flow."

Henri Moissan electrolyzed $KHF_2$ (potassium bifluoride) dissolved in liquid HF in order to isolate fluorine. For this work, he was awarded the 1906 Nobel Prize in Chemistry.

## NOTORIOUSLY REACTIVE

Fluorine's quest for electrons is reflected in its being the most electronegative of all the elements known. Electronegativity is a measure of how likely an element is to attract electrons to itself, and this tendency makes fluorine highly reactive. As a pale yellow gas, fluorine will react violently with just about anything, with only a few of the aloof noble gases being spared.

## CFCs

More trouble follows fluorine into the realm of organic chemistry, notably in the form of chlorofluorocarbons, or CFCs. In the latter part of the twentieth century, CFCs were widely used as refrigerants and propellants in aerosol cans. Their use eventually led to a public outcry when they were connected to the destruction of the ozone layer around the Earth. Ozone has a crucial role to play in protecting the Earth from the harmful ultraviolet rays of the Sun, and CFCs were proved to be depleting this shield. Their use was phased out as part of an international agreement called the Montreal Protocol, first enforced in 1989.

## HEALTHY TEETH

On a more encouraging note, fluorine has had a positive role to play in at least one area of public health. As an addition to both the water supply and to toothpastes, fluoride ions help to reverse a process known as demineralization, which causes the destruction of tooth enamel.

◀ Moissan electrolyzed $KHF_2$ (potassium bifluoride) dissolved in liquid HF in order to isolate fluorine. For this work, he was awarded the 1906 Nobel Prize in Chemistry.

▲ The ubiquitous use of chlorofluorocarbons as aerosol propellants raised concerns over the destruction of the ozone layer, which ultimately led to the phasing out of CFCs.

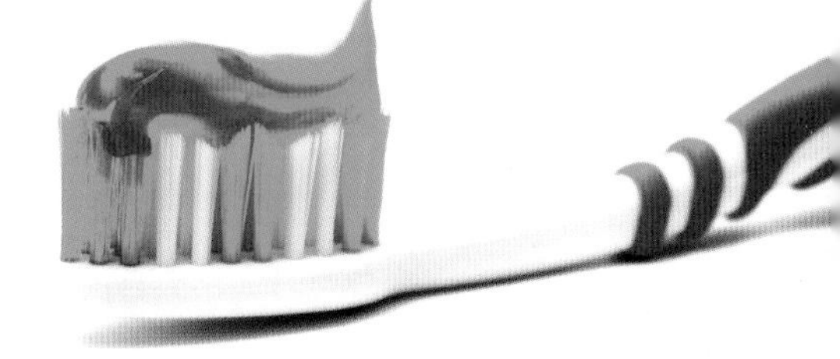

▶ Fluorine's (or, more accurately, fluoride's) use in toothpastes and water supplies helps to strengthen tooth enamel and therefore reduce tooth decay.

# Chlorine

| Chemical symbol | Cl |
|---|---|
| Atomic number | 17 |
| Atomic mass | 35.45 |
| Boiling point | −29.27°F (−34.04°C) |
| Melting point | −150.7°F (−101.5°C) |
| Electron configuration | 2.8.7 |

**Sitting below fluorine and above bromine, chlorine is another element with a checkered past of both good and evil. As both the pure diatomic (where a pair of atoms are chemically bonded to one another to make a single molecule) element, $Cl_2$, and as a component of a wide-ranging collection of compounds, chlorine has inflicted terror and saved countless lives throughout recent history.**

Discovered in 1774 by Swede Carl Wilhelm Scheele, chlorine had to wait until 1810 when Davy confirmed it as an element before its recognition was complete. Scheele had indeed made chlorine from a reaction between maganese(IV) oxide and hydrochloric acid (HCl), but he thought that he had produced another compound rather than an element.

## GAS ATTACK

In April 1915 in Ypres, France, German troops released over 150 tons of chlorine, representing the first use of a poisonous gas on the Western Front. The attack resulted in the death of approximately 5,000 French and Algerian troops. Horror stories surrounding the slow but unrelenting cloud of distinctive yellow-green gas moving toward the Allies make for some of the most harrowing reminiscences of the Great War, especially when one considers the gruesome manner in which chlorine attacks the body. Horrible choking results, and mucous membranes in the throat and lungs, eyes and skin are attacked as the chlorine reacts with water to form hypochlorous and hydrochloric acids.

## ENVIRONMENTAL AND HEALTH THREAT

Chlorine is found in several highly controversial and historically widely used compounds, including the notorious insecticide DDT (dichlorodiphenyltrichloroethane). The inventor of DDT, Paul Müller (1899–1965), won the Nobel Prize for his work, largely because DDT had been so effective. However, in 1962, the book *Silent Spring* by Rachel Carson (1907–1964) shed doubt upon the wisdom of such widespread use of chemical substances. At the time, there was no real research on the long-term effects of releasing such chemicals into the environment, and the book outlined many harmful effects of their use. Among other things, the book spurred the modern environmentalist movement, and precipitated a ban on DDT for agricultural use in the United States in 1972. CFCs (chlorofluorocarbons, the controversial refrigerants and propellants) and THMs (trihalomethanes, by-products of water chlorination) also have some gruesome stories of their own.

The chlorine-containing compound DDT was a popular ingredient of commercial insecticides in the 1960s, prior to its use being outlawed in the following decade.

During World War I, significant use of chemical weapons, including tear gas, mustard gas, and phosgene, as well as chlorine, led to soldiers being issued gas masks.

As a vital addition to drinking water, chlorine's oxidizing ability has saved millions of lives around the world in the fight against waterborne diseases.

## VITAL SAVIOR

Conversely, the chlorination of water can be considered one of humanity's greatest chemistry-based achievements. The carefully controlled introduction of chlorine into the water supply might be the single most important piece of public health action over the last 120 years. In this role, chlorine's ability to oxidize (and therefore kill) many waterborne bacteria and viruses means that billions of people have been protected from waterborne diseases such as typhoid and cholera.

# Bromine

| Chemical symbol | Br |
|---|---|
| Atomic number | 35 |
| Atomic mass | 79.904 |
| Boiling point | 137.8°F (58.8°C) |
| Melting point | 19.04°F (-7.2°C) |
| Electron configuration | 2.8.18.7 |

**Bromine is one of only two of the 118 known elements (mercury being the other) that is a liquid at room temperature. It is also one of the few elements that exist as a diatomic molecule in their elemental state. The highly reactive orange-brown nonmetal is prevalent on Earth in the bromide salts found in seawater, and one of its common compounds has a mythical connection to the control of sexual urges.**

Bromine was first extracted from brine by Antoine Jérôme Balard (1802–76) in 1826. Carl Jacob Löwig (1803–90) had discovered the element a year earlier, but Balard published first. He isolated the element by reacting it with chlorine and recognizing that the red liquid that was produced was something new and interesting. His original name for the element was "muride," but the name was changed to bromine by the French Academy of Sciences—drawn from the Greek *bromos*, meaning "stench"—in order for it to better match its recently found group 17 companions, chlorine and iodine.

### POTASSIUM BROMIDE

Bromine is a nasty, corrosive, volatile element that can prove fatal to humans in very small doses. On the other hand, bromides (salts that are created when bromine atoms gain electrons, form negative ions, and combine with positive metal atoms) have been used in various medicinal applications for centuries. Potassium bromide was first suggested as a treatment for syphilis but soon found favor as a drug used to control epilepsy and convulsions. Modern drugs such as phenobarbital have now superseded that simple salt, but, interestingly, it persists in the treatment of epilepsy in dogs. In this sedative role, potassium bromide was also promoted as a substance that could be used to suppress sexual urges, and it has a mythical reputation for being added to British soldiers' tea in World War I for that purpose!

French chemist Antoine Jérôme Balard is usually credited with the discovery of bromine, but as is sometimes the case, an additional person is co-credited—in bromine's case, the German Carl Jacob Löwig.

## QUELLING OTHER FIRES

Bromine is incorporated into chemicals used as fire-retardants. Large organobromine compounds can be used in the manufacture of furniture foams in order to increase safety. In so-called halon fire extinguishers, bromine forms part of other, smaller organic molecules such as ($CBrF_3$) in Halon 1301. In both situations, bromine works via a complex mechanism that interferes with the chain reaction combustion process.

## PEST CONTROL

Bromine's unpleasantness has been utilized in fumigants and pesticides. Historically, a number of organic compounds that contain bromine (known as organobromines) have been used in such applications, including bromomethane $CH_3Br$ (aka methyl bromide), and 1,2-dibromo-3-chloropropane $CH_2BrCHBrCH_2Cl$ (aka DBCP). Despite their effectiveness, such compounds have fallen foul of environmental regulations, and they have been largely phased out.

Bromotrifluoromethane, better known as Halon 1301, was developed at the request of the U.S. Army to replace other, more toxic fire-suppressors.

Bromides have been used in pharmaceutical preparations for centuries. These throat lozenges combined menthol with ammonium bromide, noted for its calming and hypnotic effects.

# Iodine

| Chemical symbol | I |
| --- | --- |
| Atomic number | 53 |
| Atomic mass | 126.90 |
| Boiling point | 363.92°F (184.4°C) |
| Melting point | 236.66°F (113.7°C) |
| Electron configuration | 2.8.18.18.7 |

**The largest and heaviest of the naturally occurring halogens, iodine is also the least reactive. Solid iodine exhibits a fascinating and unusual property at room temperature: it sublimes, turning directly from a solid to a gas with no liquid phase. It was the purple vapor produced as a result of this sublimation that suggested iodine's name, which derives from the Greek *iodes*, meaning "violet."**

Like bromine, the family member directly above it on the periodic table, iodine is recovered from seawater, and its discovery came in 1811 when the French chemist Bernard Courtois (1777–1838) was experimenting with seaweed. The Englishman Humphry Davy also claimed the discovery of iodine for his own, which proved to be a bit of a problem since the French and the English were at war at the time.

### DISINFECTANT

In another link to bromine, iodine has had, and continues to have, an important role to play in medicine. A traditional and familiar use for iodine is as an antiseptic. Despite being potentially damaging in large doses, the controlled application of iodine creates a very effective antiseptic barrier. As a tincture of iodine (an ethanol and water solution of the element and iodide ions), it is used topically on skin to disinfect around wounds or incisions. In this form, it leaves a familiar orange stain.

Solid iodine, which exists as blackish-gray crystals, will sublime at room temperature to produce the distinctive purple vapor that is often associated with the element.

## A BALANCING ACT

Iodine plays a crucial role in the human body and it needs to be regulated carefully to ensure correct body function. The thyroid gland in the neck is one organ where iodine accumulates, and the specific concentration of iodine there—either too high or too low—can lead to an overactive thyroid (hyperthyroidism) or an underactive thyroid (hypothyroidism). The latter may cause an abnormal swelling of the neck, known as a goiter.

## IODINE-131

The most well-known of iodine's many isotopes is perhaps iodine-131. A radioactive isotope, it decays by releasing beta particles, and it is known to accumulate in the thyroid gland in humans. In such circumstances, thyroid cancers have been observed. As a product of many nuclear reactions, it posed a significant threat to human health in the wake of both the Chernobyl disaster in 1986 and the more recent nuclear accident in Fukushima, Japan, in 2011.

**Iodide ions, often in the form of sodium iodide, are one of the ingredients in iodine tincture (along with water, iodine, and alcohol) and are added in order to help dissolve the elemental $I_2$.**

# Astatine

| | |
|---|---|
| **Chemical symbol** | At |
| **Atomic number** | 85 |
| **Atomic mass** | 210 (longest-living isotope) |
| **Boiling point** | 638.6°F (337°C) |
| **Melting point** | 575.6°F (302°C) |
| **Electron configuration** | 2.8.18.32.18.7 |

**Astatine might be considered the poster child of elusive elements. Although it is one of the elements that occur naturally in the Earth's crust, across the whole planet there is estimated to be only a few grams in total. Add the fact that all of astatine's isotopes are radioactive, most with very short half-lives, and it is hardly surprising that very little is known about the chemistry of element number 85.**

When Henry Moseley established the concept of ordering elements by their atomic number in 1913, it became apparent that seven elements were missing from the periodic table. One of these was the element with atomic number 85. This led several scientists to look for (and claim that they had found) the elusive element. But it wasn't until after World War II that a group at the University of California, Berkeley, was given credit for the element's discovery, and the period six halogen finally got its modern name.

## THE ELUSIVE ELEMENT

Astatine's name is a good indicator of its elusiveness. The Greek *astatos* means "unstable," and even though astatine is technically "naturally occurring," with so little actually in existence, that classification is somewhat academic.

**Emilio Segrè was a co-discoverer of two elements: astatine and technetium. He also co-discovered the antiproton in 1955, leading to his sharing the 1959 Nobel Prize in Physics.**

# NOBLE GASES

**All of the noble gases were discovered relatively late when compared to the bulk of the elements on the periodic table. One reason for this is their stability. If an element does not react with much, then it tends not to appear in many compounds that one might come across. Throw in the fact that the group 18 elements are colorless, odorless, and tasteless gases, and you've got some pretty-difficult-to-find elements.**

On the following pages:

| | | | |
|---|---|---|---|
| **He** | Helium | **Kr** | Krypton |
| **Ne** | Neon | **Xe** | Xenon |
| **Ar** | Argon | **Rn** | Radon |

| 1 | 2 | 3 | 4 | 5 | 6 | 7 | 8 | 9 | 10 | 11 | 12 | 13 | 14 | 15 | 16 | 17 | 18 |
|---|---|---|---|---|---|---|---|---|---|---|---|---|---|---|---|---|---|
| 1 H | | | | | | | | | | | | | | | | | 2 **He** |
| 3 Li | 4 Be | | | | | | | | | | | 5 B | 6 C | 7 N | 8 O | 9 F | 10 **Ne** |
| 11 Na | 12 Mg | | | | | | | | | | | 13 Al | 14 Si | 15 P | 16 S | 17 Cl | 18 **Ar** |
| 19 K | 20 Ca | 21 Sc | 22 Ti | 23 V | 24 Cr | 25 Mn | 26 Fe | 27 Co | 28 Ni | 29 Cu | 30 Zn | 31 Ga | 32 Ge | 33 As | 34 Se | 35 Br | 36 **Kr** |
| 37 Rb | 38 Sr | 39 Y | 40 Zr | 41 Nb | 42 Mo | 43 Tc | 44 Ru | 45 Rh | 46 Pd | 47 Ag | 48 Cd | 49 In | 50 Sn | 51 Sb | 52 Te | 53 I | 54 **Xe** |
| 55 Cs | 56 Ba | 57-71 | 72 Hf | 73 Ta | 74 W | 75 Re | 76 Os | 77 Ir | 78 Pt | 79 Au | 80 Hg | 81 Tl | 82 Pb | 83 Bi | 84 Po | 85 At | 86 **Rn** |
| 87 Fr | 88 Ra | 89-103 | 104 Rf | 105 Db | 106 Sg | 107 Bh | 108 Hs | 109 Mt | 110 Ds | 111 Rg | 112 Cn | 113 Nh | 114 Fl | 115 Mc | 116 Lv | 117 Ts | 118 **Og** |
| | | | 57 La | 58 Ce | 59 Pr | 60 Nd | 61 Pm | 62 Sm | 63 Eu | 64 Gd | 65 Tb | 66 Dy | 67 Ho | 68 Er | 69 Tm | 70 Yb | 71 Lu |
| | | | 89 Ac | 90 Th | 91 Pa | 92 U | 93 Np | 94 Pu | 95 Am | 96 Cm | 97 Bk | 98 Cf | 99 Es | 100 Fm | 101 Md | 102 No | 103 Lr |

## DOMINO EFFECT

The discovery of argon came first, in 1894, but the discovery of the bulk of the noble gases was part of a domino effect. Morris William Travers (1872–1961) and William Ramsay (1852–1916) carried out increasingly sophisticated investigations on samples of air, and discovered krypton, neon, and xenon within three months of one another between May and July 1898, a remarkably prolific and condensed run of filling in gaps in the periodic table.

Some may think of the noble gases as a bit boring, and for many years their alternative collective name, the *inert* (meaning "unreactive") gases, was an appropriate designation. It took well over half a century after their discovery before the first compound of any of them was synthesized.

## BRIGHT LIGHTS AND CITYSCAPES

Perhaps best known for their application in what are often referred to generically as "neon" lights, the ability of many of the noble gases to emit vivid colors when exposed to high voltages is due to their electrons being excited to higher energy states, and then releasing that energy as they return to their ground (unexcited) state. Each element has a unique colored fingerprint known as its spectrum, and the noble gases are particularly adept at producing the familiar city nightscapes.

Discharge tubes filled with some of the noble gases—helium, neon, argon, krypton, and xenon—emit bright, colored light as the elements' electrons are excited by high voltages.

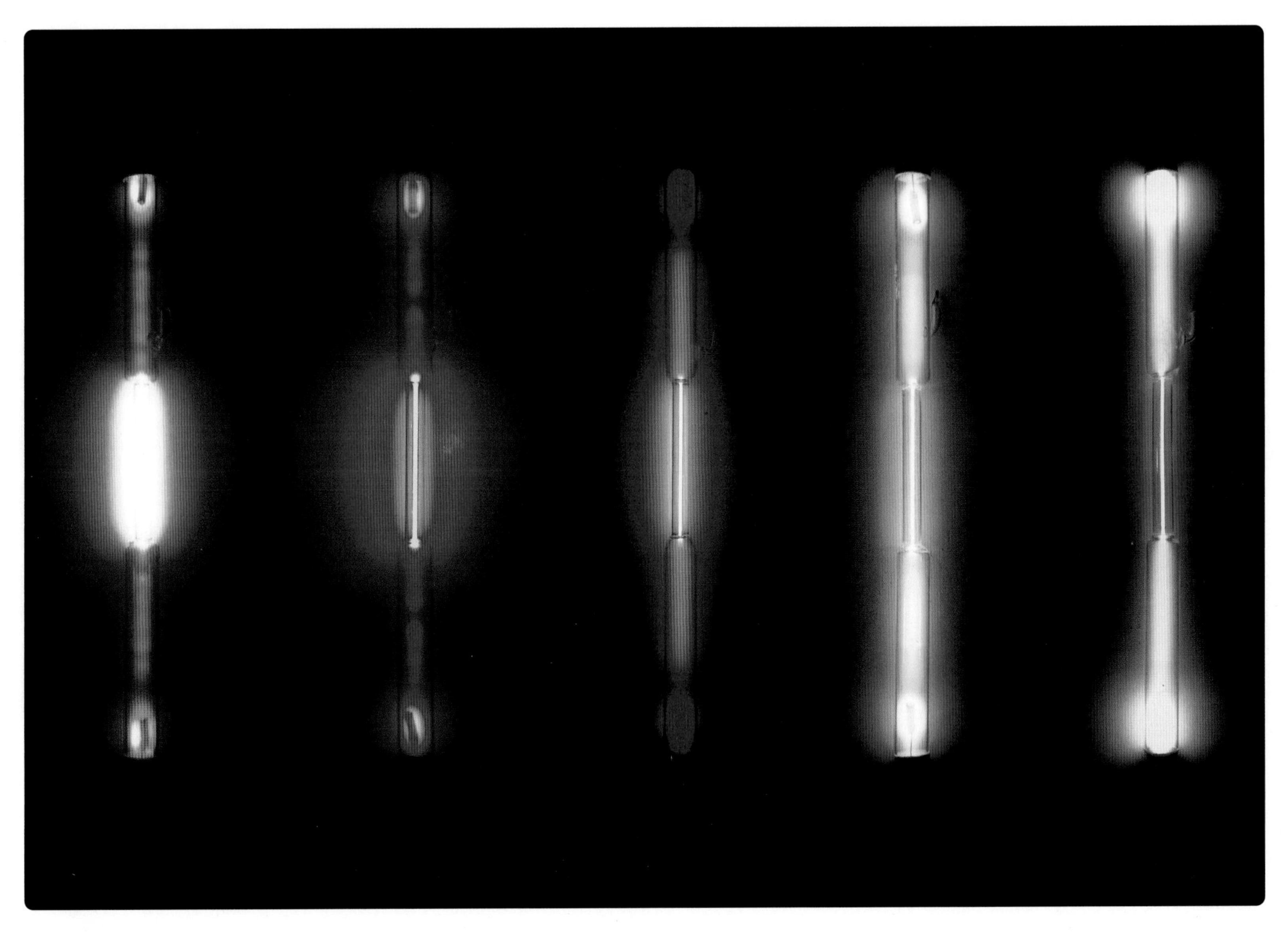

# Helium

**Chemical symbol**
He

**Atomic number**
2

**Atomic mass**
4.0026

**Boiling point**
-452.07°F (-268.92°C)

**Melting point**
-458°F (-272.2°C)

**Electron configuration**
2

**Helium is the second most abundant element in the universe, but there is precious little of it on Earth. How can this be? The clue to its whereabouts is given away by element number 2's name, which is derived from the Greek *helios*, meaning "sun." The Sun is largely a ball of burning hydrogen gas, and helium is a product of the fusion of two hydrogen atoms.**

On Earth, the uranium mineral uraninite ($UO_2$) proved to be an unlikely source of helium gas. With the observation, made during a total eclipse of the Sun, of a previously unknown spectral line in sunlight, astronomers Pierre Janssen (1824–1907) and Norman Lockyer (1836–1920) concluded that the Sun must contain an element that had not yet been identified. William Ramsay matched the data from the sunlight with data that he collected from a gas that was liberated when uraninite reacted with acid. Two scientists in Sweden, Per Teodor Cleve (1840–1905) and Nils Abraham Langlet (1868–1936), replicated Ramsay's work almost simultaneously, and sources will often credit both parties with the discovery.

## INERT

As a noble gas, helium is pretty typical of group 18 in terms of its lack of reactivity. It has very tiny atoms and as such its negative electrons are very close to its positive nucleus, making the attraction between the two strong. This leads to the highest first ionization energy (the energy required to remove an electron) of all the elements, and means that helium is generally resistant to chemical reactions.

The Sun is a star that fuses hydrogen nuclei together to produce helium. With that fusion comes the release of huge amounts of energy.

## GOING UP

As the second lightest element and as a gas, helium has found a number of uses in balloon technology. Ranging from airships to weather balloons and the more frivolous party balloons, helium's prevalent use in these applications, along with some changes in government policy in the United States, recently created some panic over a helium shortage. In the 1920s, the U.S. government established the National Helium Reserve. Originally set up to ensure a reliable source of helium for military and space exploration (helium can be used as a coolant), the program was abandoned in 1996, but not before huge reserves of the gas were released for sale, ultimately driving up prices. This was a big problem for science, since so many experiments rely upon liquid helium at a temperature of –452.2°F (–269°C) to provide the ultra-cold conditions often required for investigation at the atomic level. Some of the fear was dispelled in June 2016 when a potentially massive new source of helium gas was discovered in Tanzania.

The USS *Macon*, a helium-filled airship, flying over New York City in 1933. It acted as a "flying aircraft carrier" that smaller aircraft could be launched from.

# Neon

**Chemical symbol**
Ne

**Atomic number**
10

**Atomic mass**
20.18

**Boiling point**
-410.88°F (-246.05°C)

**Melting point**
-415.46°F (-248.59°C)

**Electron configuration**
2.8

The dazzling neon lights of a Las Vegas casino. The characteristic red glow is now synonymous with the term "neon lights."

**Like its other family members, neon is inert and relatively uncommon (it is only the eighty-second most abundant element on Earth), but it does have some important and well-known uses. "Neon" lights are how most people will be familiar with the element. In other applications, neon also finds use as a low-temperature refrigerant, with a boiling point of around -411°F (-246°C).**

## SEEK AND YE SHALL FIND

With argon (with an atomic weight of approximately 40) and helium (with an atomic weight of approximately 4) already discovered, Ramsay and Travers went looking for a gas that they thought would have a mass somewhere between the two, and that we now realize would be neon. In fact, while seeking neon they ended up discovering krypton first, in May 1898. However, they didn't have to wait long to discover the member of group 18 that they were originally seeking. They finally extracted element number 10 via the fractional distillation of a sample of argon in June 1898.

## CRIMSON BLAZE

The name given to the element was derived from the Greek *neos*, meaning "new," and was originally suggested in a slightly different form, *novum*, by Ramsay's son, Willie. With a nod to the future use of the element, Travers wrote of the spectrum that was observed in the lightest fraction of the distilled air: "The blaze of crimson light from the tube told its own story, and it was a sight to dwell upon and never to forget." Neon lights are now a familiar sight in cities all over the world, though it is only those that emit a bright red light that are likely to be utilizing the element in its purest form. Different combinations of gases generate other colors; for example, xenon and mercury vapor produce vivid blue-greens.

## STABLE ISOTOPES

Neon has three isotopes that occur naturally—neon-20, neon-21, and neon-22, with neon-20 being by far the most abundant. Like helium, neon atoms have extremely high first ionization energies, and as such, they are resistant to reaction with all known substances. The unwillingness of the isotopes to react chemically is matched by the stability of their nuclei, none of which show any radioactive behavior.

## KEY FIGURE

### WILLIAM RAMSAY 1852–1916

William Ramsay, along with Lord Rayleigh and William Travers, managed to isolate five of the seven group 18 elements over a period of roughly four years at the end of the nineteenth century. A Scotsman, Ramsay basically "invented" group 18 via his discoveries, as the quotation that accompanied his 1904 Nobel Prize in Chemistry suggests: "in recognition of his services in the discovery of the inert gaseous elements in air, and his determination of their place in the periodic system."

# Argon

**Chemical symbol**
Ar

**Atomic number**
18

**Atomic mass**
39.95

**Boiling point**
-302.53°F (-185.85°C)

**Melting point**
-308.81°F (-189.34°C)

**Electron configuration**
2.8.8

**Another colorless, odorless, and tasteless gas, argon is certainly representative of its group 18 family members. However, argon is relatively plentiful; indeed, it is the third most abundant gaseous element (behind nitrogen and oxygen) in the atmosphere.**

Because of its relative abundance, argon proved to be the easiest of the noble gases to find. It was therefore the first to be discovered, and argon's discovery is a colossally important one in the bigger story of the evolution of the periodic table.

## A WHOLE NEW GROUP

It all started with Henry Cavendish in the late eighteenth century when he was experimenting on air. He noticed that a tiny fraction of the air never reacted during his experiments, but, perhaps because the other gases that he was investigating distracted him, argon remained undiscovered. A hundred years later, Lord Rayleigh (1842–1919) was experimenting with nitrogen. Rayleigh discovered that a number of the samples of nitrogen that he took from air were heavier and more dense than the samples that he obtained from sources that did not originate from air. Rayleigh had a number of theories as to why this was, but it took collaboration with William Ramsay to finally solve the mystery. By examining the spectrum of the gas isolated from the air samples, the two concluded that in addition to nitrogen, there was a previously undiscovered element present. In 1894, Ramsay wrote to Rayleigh saying, "Has it occurred to you that there is room for gaseous elements at the end of the first column of the periodic table?" Essentially, Ramsay was suggesting that the new gas (which was named after the Greek *argos*, meaning "lazy" or "idle") could be part of a whole new group of elements that Mendeleev's early periodic table did not include.

## NONREACTIVE ATMOSPHERE

Most of argon's uses are based upon its inertness. As such, it is used in a number of applications where a nonreactive atmosphere is required, such as in electric lights and fluorescent tubes. It is also used in the welding of aluminum, where the argon protects the metal from coming into contact with oxygen and therefore prevents unwanted oxidation reactions from occurring.

Argon can be used as a "shielding gas" in welding, where the inert atmosphere that it creates prevents water vapor and oxygen gas from interfering with the welding process.

# Krypton

**Chemical symbol**
Kr

**Atomic number**
36

**Atomic mass**
83.798

**Boiling point**
-244.15°F (-153.42°C)

**Melting point**
-251.27°F (-157.37°C)

**Electron configuration**
2.8.18.8

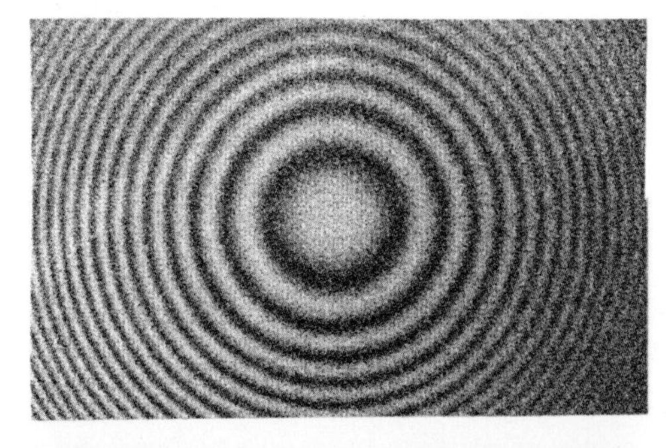

▲ Interference rings produced by a Fabry-Pérot interferometer. The lower of the two is the pattern produced by krypton-86, which was the basis for the determination of the SI base unit of length.

**Krypton exhibits the basic physical properties one would expect of a quintessential member of group 18. Like the other noble gases, because of its inertness and relative lack of abundance, krypton was not easy to find. This is reflected in its name, taken from the Greek *kryptos*, meaning "hidden."**

With argon and helium already identified as new members of the newest group, Ramsay and Travers went in search of another element that they thought might sit between those two elements. With the atomic masses of the two new gases established as approximately 4 and 40 respectively, they thought that they should look for a member of the group that had a mass of about 20. Their hunch was right, but when they went hunting for what would eventually turn out to be neon, rather than finding a lighter gas than argon, they came across an even heavier one in the process. Once again isolated through the distillation of liquid air, and once again pinpointed through a new line in the spectrum (this time green), another noble gas was identified, this time with an atomic mass between those of bromine and rubidium. This placed the new element perfectly in group 18, and in May 1898, krypton was "born."

## THE LENGTH OF A METER

Krypton's uses are very similar to those of the other noble gases; that is, it is used in lighting applications and in places where inert atmospheres are required. But krypton also has a special place in the history of the International System of Units, or SI. The base unit for length as defined by the SI is the meter. Like all such base units, the base quantity must be defined in a very specific way. Today, the meter is defined as the distance traveled by light in 1/299,792,458 of a second, but it has been characterized in several other ways in the past. One such definition involved a particular isotope of krypton: krypton-86. Between 1960 and 1983, the SI base unit of length was defined as: "1,650,763.73 wavelengths of the orange-red emission line in the electromagnetic spectrum of the krypton-86 atom in a vacuum."

Krypton has a dizzying array of isotopes, most radioactive, which have masses ranging in the high 60s all the way through to those with masses of over 100. The isotopes have an equally wide range of radioactive properties as well, but krypton-86 is very stable. It was this stability that helped cement its use in the definition of the meter.

▼ The international prototype meter bar, the original standard of length prior to 1960, when the use of krypton-86 replaced it.

# Xenon

**Chemical symbol**
Xe

**Atomic number**
54

**Atomic mass**
131.29

**Boiling point**
-162.58°F (-108.1°C)

**Melting point**
-169.15°F (-111.75°C)

**Electron configuration**
2.8.18.18.8

**Element number 54 occupies the final position in period 5. Like the other members of group 18, xenon is a colorless and odorless gas, and that means it is far from easy to find. Its name is taken from the Greek *xenos*, meaning "stranger."**

### THE FINAL GAS

With argon, neon, krypton, and helium all in the bag, xenon was the final noble gas waiting to be discovered. In July 1898, Ramsay and Travers continued their work on the distillation of liquid air, this time with new equipment given to them by the German/British chemist Ludwig Mond (1839–1909). After continually extracting fractions from a sample of krypton, they finally managed to isolate element number 54.

### LATE TO THE PERIODIC PARTY

Much like the other noble gases, xenon's chemistry is limited when compared to many other elements, but is important simply for the fact that it helped to shatter the perceived (and, at the time, real) aloofness of the group 18 elements. In 1962, British-born chemist Neil Bartlett (1932–2008) made the first compound of any of the noble gases. Bartlett synthesized a compound of xenon, platinum, and fluorine, by reacting platinum hexafluoride with xenon. The formation of the first compound to contain a noble gas was a seminal moment, since it opened up a whole new branch of chemistry. Bartlett's work led to more compounds of xenon with fluorine and oxygen, such as $XeF_2$, $XeF_4$, $XeO_2F_2$, $XeO_3$, $XeO_2$, and $XeF_6$. Other noble-gas compounds have been made, notably using krypton, but xenon is by far the most interesting of the noble gases in terms of the compounds that it can form.

### ION PROPULSION

Xenon as an element has been used as a general anesthetic, as an ion engine by NASA for spacecraft propulsion, and as a source of light. Here, the movement of electrons within excited xenon atoms emits a characteristic blue color. In ion engines, the propellant (in this case xenon gas) is ionized (has electrons removed) and then the resultant positively charged species are accelerated to speeds of tens of thousands of miles per hour using an electrical field. In turn, this produces thrust. Xenon has a host of chemical and physical properties that make it suitable for engines utilized by some spacecraft: It is inert, so will not cause corrosion; as a gas, it can be compressed, so a large quantity of it can be stored in a very small space; it has a relatively low ionization energy to allow for easy ionization; and it has a high atomic mass, which creates a significant thrust.

Longer-lasting, brighter, and capable of producing more intense light than conventional headlights, xenon lighting is now being used as a standard in many high-end vehicle models.

# Radon

**Chemical symbol**
Rn

**Atomic number**
86

**Atomic mass**
222 (longest-living isotope)

**Boiling point**
-79.1°F (-61.7°C)

**Melting point**
-96°F (-71°C)

**Electron configuration**
2.8.18.32.18.8

**Until recent discovery and confirmation of element 118, oganesson, radon was officially the heaviest and most dangerous of the noble gases. Although no longer the heaviest, because of its potent radioactivity and the fact that oganesson remains a laboratory-based curiosity, radon's reputation as the most fearsome member of group 18 is likely to continue.**

## CAUSING CONFUSION

Right at the end of the nineteenth century, the understanding of the phenomenon that we now know as radiation was in its infancy. One of the consequences of this limited understanding was that radon's birth as an element was a confusing one. Originally recognized as a gas that was given off from the element radium, it was given the literal name "radium emanation." Things got very muddled, very quickly, when similar "emanations" were observed from the elements thorium and actinium, and were given names such as thoron and actinon, as well as thorium emanation and actinium emanation. When Friedrich Dorn (1848–1916) was credited with the official discovery of element number 86 in 1900, there was still a great deal of confusion surrounding what it should be called. All this, combined with the fact that these atoms were all radioactive with short half-lives, meant that confusion was high! In the early part of the twentieth century, many periodic tables carried the symbol Em, for emanation, and the king of the noble gases, William Ramsay, had also proposed the name "niton" for number 86. It was not until the mid-1920s that radon became the accepted name.

The room for preparing radon at the Radium Institute (now the Curie Institute) in Paris, France. The institute was originally founded as a laboratory for Marie Curie in 1909.

## HEALTH RISK

The radioactive nature of element number 86 is immortalized in its name, which comes from element number 88, radium, which in turn comes from the Latin *radius*, meaning "ray." During the final two decades of the twentieth century, radon gas was recognized as a potentially serious health risk. Naturally occurring radon was found to have accumulated in certain locations, and in 1988 the International Agency for Research on Cancer declared it a human carcinogen.

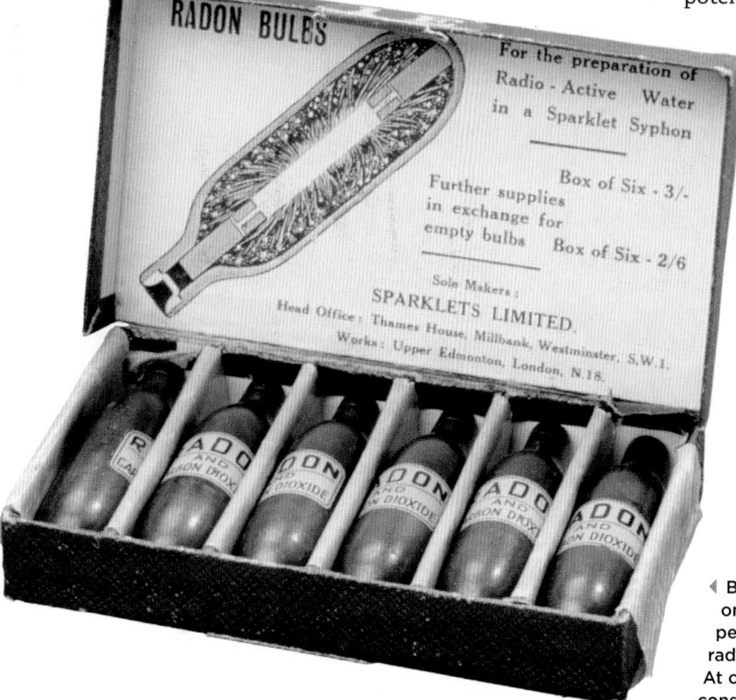



Bulbs filled with radon were once offered for sale so that people could produce their own radon-infused water at home. At one time, such water was considered a health tonic.

# Index

# Picture credits and author biography

The publisher would like to thank the following for permission to reproduce copyright material:

**Alamy**
D. Hurst 67 (bottom right); Eraza Collection 26 (right); Imagedoc 73 (top); INTERFOTO 40 (left), 49; Everett Collection Historical 117 (bottom); Science Photo Library 36 (top); sciencephotos 109; Zoonar GmbH 79 (bottom); ZUMA Press, Inc. 81 (bottom).

**Creative Commons**
AkselA 78 (bottom); Alchemist-hp 82 (right), 119; Alexander C. Wimmer 53 (bottom); Andrew Dunn/www.andrewdunnphoto.com 66 (bottom); Arma95 27 (right); CERN 44 (bottom); DanCentury 38 (bottom); Danny S. 116 (right); David Monniaux 53 (top); Dennis "S. K" (Dnn87) 13 (top); Energy.gov 91 (bottom); ESA/Hubble 6; http://images-of-elements.com 66 (top), 83 (bottom), 107 (top); Jochen Gschnaller 99 (left); Kristian Molhave 47 (top); Krizu 42 (left); Metalle-w 46 (left); NobbiP 44 (top); Ondrej Martin Mach57; Rob Lavinsky, iRocks.com 73 (bottom), 74 (right); Sailko 55 (right); Sam L. 91 (top); Spidey71 79 (top); Stan Shebs 107 (bottom); Steve Jervetson 11 (top); Thomas Nguyen 63 (top); Yale Rosen/Dr. Mark Wick 20 (top).

**Getty Images**
De Agostini Picture Library 106 (left); Natasja Weitsz 75; Peter Ginter 42 (right); © Philip Evans 14; traveler1116 12.

**Science Photo Library**
American Institute of Physics 17 (bottom); Andrew Brookes/National Physical Laboratory 86 (bottom); Biophoto Associates 50 (top); Claude Nuridsany & Marie Perennou 116 (left); Dr. P. Marazzi 55 (left); Emilio Segre Visual Archives/American Institute of Physics 16 (top); GIPhotoStock 63 (bottom); Iowa State University/American Institute of Physics 84 (left); Mondadori Portfolio 114; National Physical Laboratory © Crown Copyright 17 (top), 123 (left, top and bottom); Patrick Landmann/Science Photo Library 94 (bottom); Philippe Plailly 47 (bottom); Public Health England 125 (bottom); SCIENCE PHOTO LIBRARY 82 (left).

**Shutterstock**
5 Second Studio 48 (right); AkeSak 83 (top); Albert Russ 43 (top), 58 (top); AlenKadr 62 (left); Alexey Kamenskiy 37 (bottom); AlteredR 80 (left); Baciu 33 (top); bluecrayola 36 (bottom); bonchan 45 (top right); Comaniciu Dan 99 (right); Coprid 38 (top); Daniel D Malone 37 (top); Denis Radovanovic 34 (bottom); design56 11 (bottom); dibrova 39; Don Pablo 32; Dsmsoft 71 (top); dumbell619 84 (right); elina 98 (right); Evannovostro 52 (left); Everett Historical 7, 9, 26 (left), 59, 93 (top), 94 (top); Film Factory 77; Gary James Calder 13 (bottom); Greentea Latte 51 (top); ilolab 105 (bottom); Imageman 33 (bottom); Imfoto 10; In Green 48 (left); Jason Dudley 52 (top); jiangdi 117 (top); Jiri Vaclavek 67 (bottom left); Joseph Sohm 111 (left); Joy Tasa 101 (top); Kirill Smirnov 55 (center); kyokyo 64; MarcelClemens 58 (bottom); Margrit Hirsch 93 (bottom); Mau Horng 111 (right); Mike Flippo 115 (right); milezaway 67 (top); Mozakim 71 (bottom); Nada B 51 (bottom); Namthip Muanthongthae 29; Nastya22 20 (bottom); Neil Lang 121 (top); NikolayN 80 (right); NV77 46 (right); Oscity 15; Piotr Zajc 30; Posonskyi Andrey 122; posteriori 113 (bottom); pryzmat 65 (bottom); Rat007 97; revers 35; RHJPhotoandillustration 40; Richard Thornton 70 (right); S-F 50 (bottom); Sarin Kunthong 41; sebra 43 (bottom left); SIAATH 106 (right); Southern Vadim 65 (top); SpaceKris 89; studio23 103 (top); Sunny_Images 124; Suttha Burawonk 85 (top); Syda Productions 86 (top); tcsaba 61 (bottom); testing 101 (bottom); Tethys Imaging LLC 21 (bottom right); thailoei92 25; Ti Santi 54 (left); Tortoon 23; totojang1977 19; Tushchakorn 31 (top); Vladimir Mucibabic 22; wacomka 34 (top); watin 98 (left); wk1003mike 103 (bottom); yanisa nithichananthorn 61 (top); Yulia Grigoryeva 62 (right).

**Wellcome Library, London**
56, 70 (left), 72 (left), 72 (right), 74 (left), 95 (left), 100, 102, 104, 110, 113 (top), 115 (left), 121 (box), 125 (top).

**Misc**
NIST/JILA/CU-Boulder 16 (bottom).

While all reasonable efforts have been made to credit photographers, the publisher apologizes for any omissions or errors, and would be pleased to make the appropriate correction in future editions of the book.

## ABOUT THE AUTHOR

Adrian Dingle is the author of numerous chemistry books for all ages, from fun books for children, such as *Awesome Chemistry Experiments for Kids*, to study guides and the *DK Encyclopedia of Science*. His books have won awards including the American Institute of Physics Communication Award and the School Library Association's Information Book Award. He is also a regular contributor to online chemistry projects and journals such as Nature Chemistry, Shmoop, The Discovery Channel and ChemMatters magazine.